Building a Resilient Digital Future

Building a Resilient Digital Future

A Comprehensive Guide to Cyber Risk Monitoring

Elizabeth Stephens

ISBN 979-8-218-42387-2 (Paperback)
ISBN 979-8-218-47687-8 (E-book)

Editing by Elizabeth Stephens
Cover art by Danièle Henderson
Book design by Elizabeth Stephens

Printed in United States of America

First edition August 2024

Visit https://dbscyber.com

Table of Contents

Figures

Forward

A Champion for Digital Security in a Time of Growing Threats

Have you ever encountered someone with an exceptional drive to make a substantial impact?

Meet Elizabeth and her dedication to cybersecurity is truly inspiring.

When we first met, I could not have predicted the profound influence she would have on my perspective. Back then, we were simply project managers collaborating on a work initiative.

Elizabeth embodies strength, perseverance, and an unwavering commitment to ethical conduct. Her decision to embark on this mission to equip everyone with the tools to safeguard themselves, their loved ones, and their professional associates, along with the digital environments they work in, from the rising tide of cyberattacks, comes as no surprise.

In our technology-driven world, the threat of cyberattacks is an increasingly critical concern. These attacks can have devastating consequences, from ransomware that paralyzes essential services to data breaches that expose sensitive information. This book, titled *Building a Resilient Digital Future: A Comprehensive Guide to Cyber Risk Monitoring*, equips you with the knowledge to navigate these challenges and establish a more secure digital environment.

This book introduces a groundbreaking approach to cybersecurity. Rather than merely responding to threats as they emerge, it advocates for a preventative approach. This entails anticipating and mitigating risks before they materialize. By leveraging principles gleaned from military intelligence, the book demonstrates how to transform data into actionable insights that can safeguard your organization from ever-evolving threats.

The human element of cybersecurity is another area of emphasis in this book. It acknowledges that people can function as both a powerful defense and a potential vulnerability. You will gain the knowledge necessary to cultivate a culture of cyber awareness within your organization, empowering your employees to share the responsibility of maintaining information security.

The digital age presents unparalleled opportunities for innovation, but it also introduces substantial risks. Cyber threats such as ransomware, phishing scams, and supply chain breaches can target any organization, jeopardizing essential assets, operations, and trust. With the projected cost of cybercrime exceeding $10 trillion annually by 2025, the need for robust digital defense mechanisms has never been greater.

This book serves as a guide for developing a Cyber Risk Monitoring Plan (CRMP) and establishing a proactive, intelligence-driven defense strategy. You'll delve into core principles outlined in "The Cyber Risk Intelligence Manifesto," gaining a deeper understanding of how to assess threats, analyze vulnerabilities, and leverage intelligence techniques to strengthen your cybersecurity posture.

The insights and examples provided in this book will equip you to safeguard your organization within the ever-shifting landscape of cyber threats. By incorporating the principles outlined in the Cyber Risk Manifesto, this guide empowers you to develop a proactive defense strategy that ensures success in the digital age.

Anne Marie Otanez
Founder and CEO of Own Your Power with Anne Marie
annemarieotanez.com

Preface

In an age where our lives are increasingly intertwined with the digital realm, the threat of cyberattacks looms larger than ever, with devastating consequences for individuals, businesses, and critical infrastructure. Amidst this escalating danger, "Building a Resilient Digital Future: A Comprehensive Guide to Cyber Risk Monitoring" emerges as your roadmap to a safer, more secure digital existence.

Within these pages, we challenge traditional, reactive cybersecurity approaches that focus on isolated threats. Instead, we embrace a holistic, systems thinking approach, viewing your digital infrastructure as a complex ecosystem of interconnected people, processes, and technologies. Inspired by military intelligence principles, this approach enables proactive risk management by understanding how various system components interact. By identifying vulnerabilities before they are exploited, you can build a more resilient defense against the ever-evolving threat landscape.

This book is not just a technical manual; it is a guide to a paradigm shift in cybersecurity. You will learn to leverage intelligence-driven strategies to anticipate and mitigate risks before they materialize, transforming raw data into actionable insights that fortify your organization. We'll also delve into the human element, empowering your employees and fostering a culture of shared responsibility for security.

Whether you're a seasoned cybersecurity professional or a business leader seeking to safeguard your organization's digital assets, this book offers practical guidance and a forward-thinking approach. You'll gain a comprehensive understanding of cyber risks, develop a tailored risk monitoring plan, and master the tools and techniques necessary to build a resilient digital future.

This is more than just a guide; it's a call to action. It's time to move beyond fear and embrace a proactive, intelligence-led approach to cybersecurity. The digital age demands nothing less.

Introduction

The digital era is a double-edged sword – offering unprecedented opportunities for innovation while exposing us to a dangerous and volatile risk landscape. Ransomware attacks cripple businesses and hospitals, phishing scams exploit human trust, and supply chain breaches allow attackers to sneak in undetected. These sophisticated threats target organizations of all sizes, jeopardizing critical assets, operations, and trust. The projected cost of cybercrime, expected to exceed $10 trillion annually by 2025, underscores the critical need for robust digital defense mechanisms.

This book is a guide to developing a Cyber Risk Monitoring Plan (CRMP) and a practical application of the principles outlined in "The Cyber Risk Intelligence Manifesto," a blueprint for building a proactive, intelligence-driven defense strategy. We will delve into the principles outlined in This manifesto provides a blueprint for building a proactive, intelligence-driven defense strategy to safeguard your organization in the face of these growing threats.

You will learn about:

1. **The Foundations of Cyber Risk Intelligence:** Discover the types of cyber threats prevalent today – from ransomware and phishing to supply chain attacks – and the importance of proactive risk management, and the intelligence techniques that empower informed decision-making.

2. **Developing Your CRMP:** Follow a detailed methodology for creating a plan tailored to your organization. Assess your cybersecurity posture, analyze risks, and establish clear mitigation strategies.

3. **Strategic Defense:** Gain practical knowledge about the technologies, tools, and best practices essential for combating both common and advanced cyber threats.

4. **Managing and Enhancing Your CRMP:** Understand how to define roles, implement training, and integrate the principles of continuous improvement to ensure your plan remains effective against evolving threats.

5. **The Future of Cyber Risk Monitoring:** Explore emerging trends, the vital role of adaptability, and how to build a resilient digital future for your organization.

This book's insights, examples, and forward-thinking approach empower you to protect your organization within the ever-changing cyber threat landscape. It leverages the Cyber Risk Manifesto, Part 0 of this guide, to forge a proactive defense that ensures your success in the digital age.

"To survive in this environment, we need more than just defenses; we need a new mindset."

"Cybersecurity risk management will always fail unless it is proactive and informed by intelligence (the discipline)."

The Cyber Risk Manifesto: Reborn

*"In an age where cyber threats evolve at an unprecedented pace, the traditional reactive approach to cybersecurity is no longer sufficient. This manifesto outlines the tenets of **intelligent cyber defense**, a proactive, data-driven strategy that empowers organizations to anticipate, identify, and mitigate risks before they can be exploited. By harnessing cutting-edge technologies, integrating risk management principles, and fostering a culture of cyber awareness, organizations can build a resilient digital fortress capable of withstanding the ever-changing threat landscape. This manifesto serves as a call to action for leaders at all levels to embrace this holistic approach, ensuring the protection of critical assets, infrastructure, and the very foundation upon which our digital society thrives."*

The relentless evolution of cyber threats demands a proactive, intelligence-led response. This manifesto outlines the tenets that will form the bedrock of modern digital defense, empowering organizations to outmaneuver adversaries and protect their critical assets.

People: Empowering the Human Element

In today's digital era, cybersecurity is not solely a technical challenge; it's a human one. People, from frontline employees to executive leadership, are both the first line of defense and the most significant potential vulnerability.

To build a resilient digital future, we must:

1. **Cultivate a Culture of Cyber Awareness:** Empower individuals at all levels with the knowledge and tools to recognize and respond to threats. This includes ongoing training on secure practices, phishing awareness, incident reporting procedures, and participation in tabletop exercises to simulate real-world scenarios.

2. **Foster Shared Responsibility:** Cybersecurity is everyone's responsibility. Break down silos and foster collaboration between IT, security teams, and business units to ensure a unified approach to risk management.

3. **Uphold Ethical Practices:** Respect user privacy, protect personal data, and avoid actions that could cause unintended harm. Ethical behavior builds trust and reinforces the integrity of our digital spaces.

Process: The Framework for Intelligent Defense

The increasing interconnectedness of our world and the sophistication of cyber threats underscore the need for a proactive Cyber Risk Monitoring Plan (CRMP). This plan is not merely a component of an organization's defense strategy; it's the backbone of its survival and resilience. To achieve this, we must:

1. **Embrace Intelligence-Driven Actions:** Transform data into actionable intelligence through collection, analysis, and application. This knowledge allows us to anticipate threats, understand adversaries' tactics, and make informed decisions that protect our assets.

2. **Prioritize Proactive Risk Management:** Anticipate potential threats and vulnerabilities to devise strategies that prevent cyber risks before they materialize. This includes regular risk assessments, vulnerability scanning, continuous monitoring of systems and networks, and utilizing tools like tabletop exercises to test response capabilities.

3. **Adopt Agile and Adaptive Operations:** The cyber threat landscape is constantly evolving. Organizations must be agile and adaptable, ready to pivot in response to new threats and vulnerabilities. This includes having pre-defined incident response playbooks and conducting regular security exercises like tabletop drills to test and refine processes.

To get started with tabletop exercises, download our free sample scenario and facilitator's guide at https://rb.gy/czt93z.

Technology: The Tools to Fortify Our Defenses

Technology is an essential enabler in the fight against cyber threats, providing us with the tools to detect, prevent, and respond to attacks. To build a resilient digital future, we must:

1. **Implement Advanced Monitoring Systems:** Leverage artificial intelligence and machine learning to predict and identify potential risks before they materialize. This includes deploying next-generation firewalls, intrusion detection systems, and encryption protocols.

2. **Adopt a Zero Trust Model:** Assume that no user or device is inherently trustworthy. Implement strict access controls, multi-factor authentication, and continuous

monitoring to verify identities and protect sensitive data.

3. **Foster Collaboration and Innovation:** Cyber threats transcend organizational and national boundaries. By sharing threat intelligence, best practices, and resources, we can collectively enhance our defenses and stay ahead of evolving threats. We must also encourage innovation in cyber defense technologies and methodologies to ensure our tools remain effective.

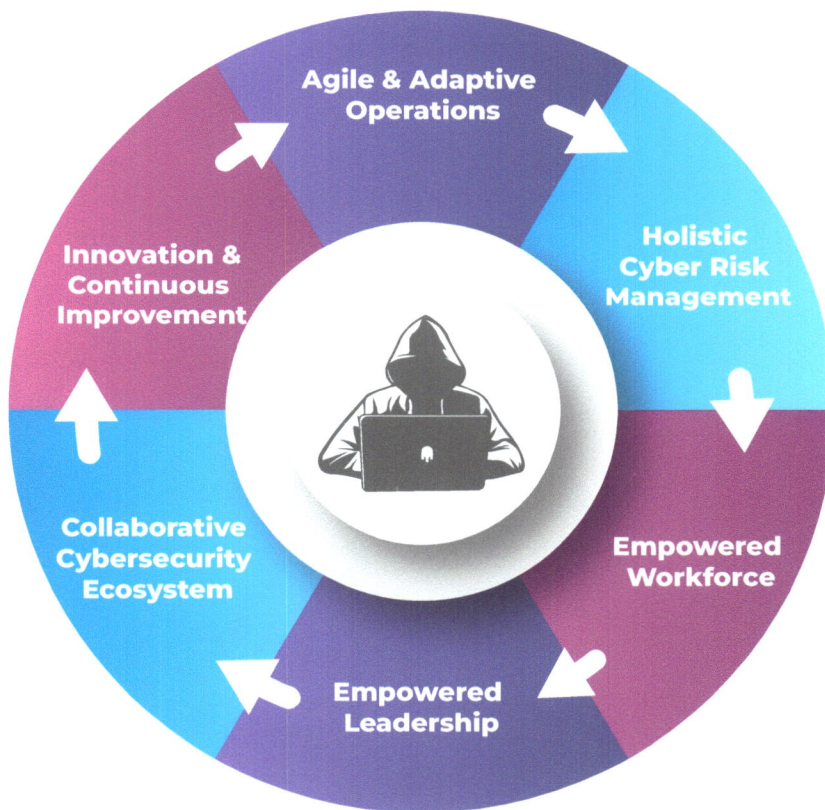

Figure 1: The Cycle of Intelligent Cyber Defense

Conclusion

This manifesto is a call to action for leadership at every level to embrace a holistic approach to cyber defense, recognizing the critical interplay of people, processes, and technology. Through proactive risk management, intelligence-driven action, and unwavering collaboration, we can build a more resilient and secure digital future for all. The time for complacency is over. The era of intelligent cyber defense is now.

Intelligence Operations Cycle: The Framework for Intelligent Defense

The Intelligence Operations Cycle is the cornerstone of effective cyber risk management. It provides a structured, systematic approach to gathering, analyzing, and applying information to inform decision-making and guide proactive defense strategies. The cycle consists of five interconnected stages, each playing a crucial role in transforming raw data into actionable intelligence.

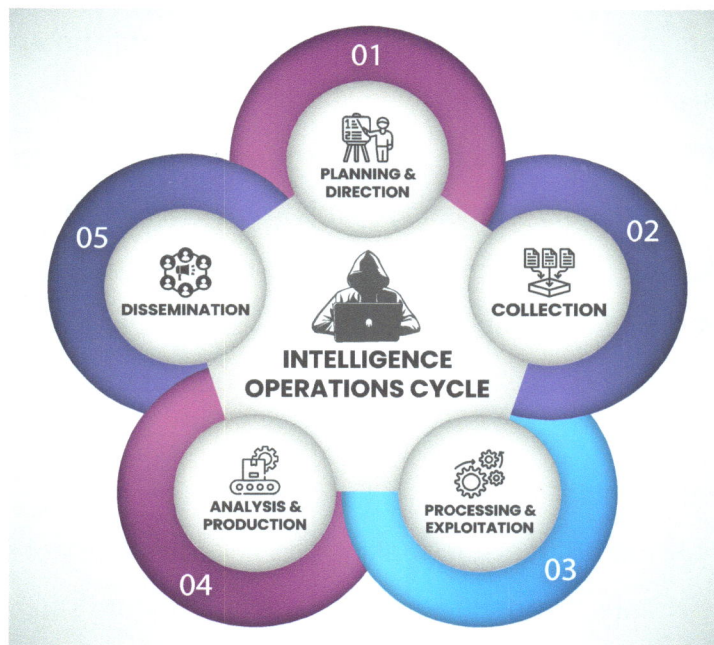

Figure 2: The Intelligence Cycle in Cyber Risk Monitoring

1. **Planning and Direction:** In this stage, the goals and objectives of the Cyber Risk Monitoring Plan are established, and the necessary resources and capabilities are identified. This stage involves defining the scope of the plan, identifying the stakeholders and their requirements, and establishing the key performance indicators (KPIs) to measure the effectiveness of the plan.

2. **Collection:** In this stage, data is collected from various sources, such as internal and external networks, social media, and threat intelligence feeds. This stage involves collecting data on potential threats, vulnerabilities, and indicators of compromise (IOCs) that could impact the organization.

3. **Processing and Exploitation:** In this stage, the collected data is processed and analyzed to identify patterns and potential threats. This stage involves applying analytics and data mining techniques to identify correlations, anomalies, and other indicators of potential threats. The goal is to identify potential threats early to prevent or mitigate their impact.

4. **Analysis and Production:** In this stage, the processed data is analyzed and translated into actionable intelligence. This stage involves analyzing the potential impact of identified threats and vulnerabilities on the organization and providing recommendations for risk mitigation or remediation. The goal is to provide decision-makers with timely and relevant information to support their decision-making processes.

5. **Dissemination:** In this stage, the intelligence is disseminated to relevant stakeholders to support decision-making and risk management activities. This stage involves sharing the intelligence with the appropriate stakeholders, such as the incident response team, management, and other key stakeholders.

Conclusion

The Intelligence Operations Cycle is a powerful methodology for managing cyber risks. By providing a structured approach to managing cyber risks, it enables organizations to identify potential threats and vulnerabilities early and make informed decisions based on timely and relevant intelligence. This methodology provides a continuous feedback loop that allows organizations to continually improve their risk management processes and capabilities.

Intelligence Operations in a Business

The digital and physical worlds are deeply interconnected, especially in critical infrastructure like data centers. These facilities house the essential servers, storage, networking equipment, and other systems that power modern businesses and governments. Their security and resilience are paramount, as disruptions can have far-reaching consequences.

Data centers are prime targets for both cyber and physical threats due to the concentration of valuable assets and interconnected systems. Cyberattacks can range from malware and ransomware to sophisticated hacking attempts, while physical threats include fire, flood, power outages, and unauthorized access. Effective cyber risk management requires addressing both the digital and physical vulnerabilities of data centers.

The key components include:

- **Servers:** Powerful computers that run applications, process data, and provide services.
- **Storage Systems:** Arrays of hard drives, solid-state drives, or other storage media that hold the data used by the servers.
- **Networking Equipment:** Routers, switches, and firewalls that enable communication within the Data Center and connect it to the outside world.
- **Power and Cooling Systems:** Uninterruptible power supplies (UPS), backup generators, and sophisticated cooling systems to ensure continuous operation and prevent equipment from overheating.
- **Physical Security:** Fences, guards, access controls, and surveillance systems to protect the facility and the valuable data within.

Types of Data Centers include:

- **Enterprise Data Centers:** Owned and operated by a single organization for their internal use.

- **Colocation Data Centers:** Provide space and infrastructure for rent to multiple customers.

- **Cloud Data Centers:** Large-scale facilities operated by cloud service providers like Amazon (AWS), Microsoft (Azure), and Google (GCP).

- **Hyperscale Data Centers:** Massive Data Centers specifically designed to support the needs of cloud computing giants.

Developing Intelligence

Intelligence is a body of knowledge collected and analyzed for specific purposes. Intelligence, in the context of cybersecurity, refers to the collection, analysis, and interpretation of data to understand the threat landscape, identify vulnerabilities, and inform decision-making. It involves turning raw data into actionable insights that enable organizations to make informed choices about how to protect their assets and operations.

There are two primary objectives of intelligence in cybersecurity:

1. **Cybersecurity Intelligence:** This focuses on providing accurate, timely, and relevant information about cyber threats, vulnerabilities, and the broader technological landscape. Its goal is to support decision-making by reducing uncertainty about an organization's cybersecurity posture.

2. **Security Intelligence (with an emphasis on Supply Chain):** This focuses on proactively identifying and mitigating risks within the supply chain, including vulnerabilities in third-party vendors, software, and hardware. It aims to safeguard the organization from external threats by building resilience into the supply chain.

By combining both types of intelligence, organizations can develop a comprehensive understanding of the risks they face and implement effective mitigation strategies.

Intelligence Cycle: The process of developing intelligence involves several stages:

1. **Planning and Direction:** Defining objectives and identifying information needs.
2. **Collection:** Gathering data from various sources, including internal and external networks, social media, and threat intelligence feeds.
3. **Processing and Exploitation:** Transforming raw data into usable information through filtering, organizing, and translating.

4. **Analysis and Production:** Analyzing processed information to identify patterns, anomalies, and potential threats, and producing actionable intelligence reports.

5. **Dissemination:** Sharing intelligence with relevant stakeholders in a timely and understandable manner.

Figure 3: Intelligence Functions

This cyclical process allows organizations to continuously refine their intelligence operations and adapt to the ever-changing threat landscape.

Intelligence operations support decision-making by:

- Describing the operational environment.

- Identifying key factors that can influence operations.

- Establishing and maintaining a multi-domain baseline of the operational environment.

- Defining and evaluating threat capabilities.

- Identifying adversaries' centers of gravity (COG) and critical vulnerabilities.

- Assessing adversaries' intentions.

The knowledge and resulting artifacts, when rationalized, are the commoditized information requirements that become the business logic for the automation of the process and analysis.

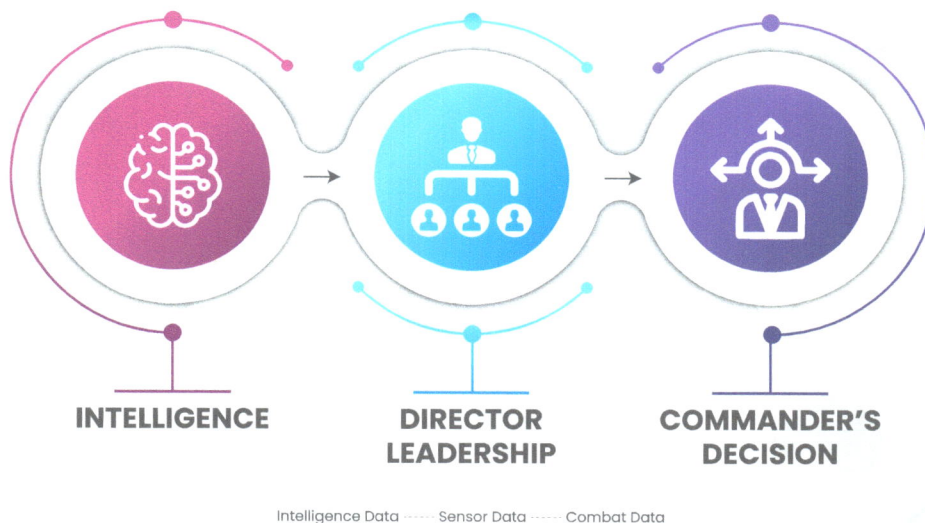

INTELLIGENCE   DIRECTOR   COMMANDER'S
                      LEADERSHIP   DECISION

Intelligence Data ····· Sensor Data ····· Combat Data

Figure 4: The Intelligence Process: Turning Data into Actionable Insights

*Information in this context is facts provided or learned about someone or something.

To be considered intelligence, data must be focused and placed in context. The organization combines business knowledge with knowledge of both the friendly and adversary situations, employing experience, judgment, and intuition to gain a complete understanding. This understanding is then applied to make informed decisions through actionable intelligence, which is the output of a process that converts data and information into knowledge tailored to specific operational decisions.

The Organization combines Business knowledge with the of the friendly and adversary situation. The Business then employs experience, judgment, and intuition to understand the complete situation.

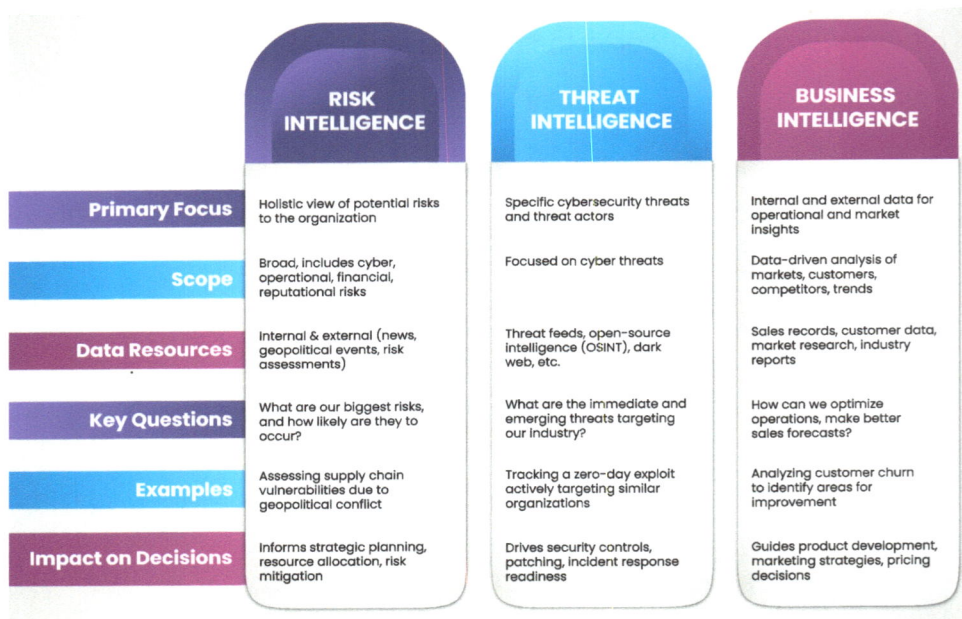

	RISK INTELLIGENCE	THREAT INTELLIGENCE	BUSINESS INTELLIGENCE
Primary Focus	Holistic view of potential risks to the organization	Specific cybersecurity threats and threat actors	Internal and external data for operational and market insights
Scope	Broad, includes cyber, operational, financial, reputational risks	Focused on cyber threats	Data-driven analysis of markets, customers, competitors, trends
Data Resources	Internal & external (news, geopolitical events, risk assessments)	Threat feeds, open-source intelligence (OSINT), dark web, etc.	Sales records, customer data, market research, industry reports
Key Questions	What are our biggest risks, and how likely are they to occur?	What are the immediate and emerging threats targeting our industry?	How can we optimize operations, make better sales forecasts?
Examples	Assessing supply chain vulnerabilities due to geopolitical conflict	Tracking a zero-day exploit actively targeting similar organizations	Analyzing customer churn to identify areas for improvement
Impact on Decisions	Informs strategic planning, resource allocation, risk mitigation	Drives security controls, patching, incident response readiness	Guides product development, marketing strategies, pricing decisions

Figure 5: The Interplay of Intelligence in Cyber Risk Management: Risk, Threat, and Business Perspectives

The organization then applies this understanding to make decisions through actionable intelligence. Intelligence is the output of a process to convert data and information into knowledge tailored to a specific operational decision.

What is Intelligence Production and Analysis?

Intelligence Production and Analysis is the filtering, recording, evaluating, and analyzing of information, and product preparation of developed intelligence.

Analysis is a process that involves sifting and sorting evaluated information to isolate significant elements related to the mission of the command, determining the significance of the information relative to information and intelligence already known, and drawing deductions about the probable meaning of the evaluated information.

Production is the conversion of evaluated material or analyzed information into intelligence either through assessment or summary.

All sources of information are integrated, analyzed, evaluated, and interpreted to prepare intelligence products or all source intelligence in support of known or anticipated user requirements.

Production or the process of analysis and synthesis is the most important action in developing usable intelligence for organization. Production, the fourth step in the intelligence cycle, helps forecast the effect gathered intelligence will have on organization's ability to accomplish the mission.

Information Cycle:

The information cycle consists of a series of related activities that translate the need for information about a particular aspect of the operational environment or threat into a knowledge-based product the commander uses in the decision-making process.

The process used to develop information is called the intelligence cycle (planning and direction, collection processing and exploitation, analysis and production, dissemination, and utilization). This cycle is a series of actions by which required information is <u>found, obtained, assembled, converted through analysis</u> <u>into intelligence</u> that is provided to decision makers, and used to inform decisions.

This means that when a piece of the puzzle is received, it is processed and then added to other pieces to become a full image.

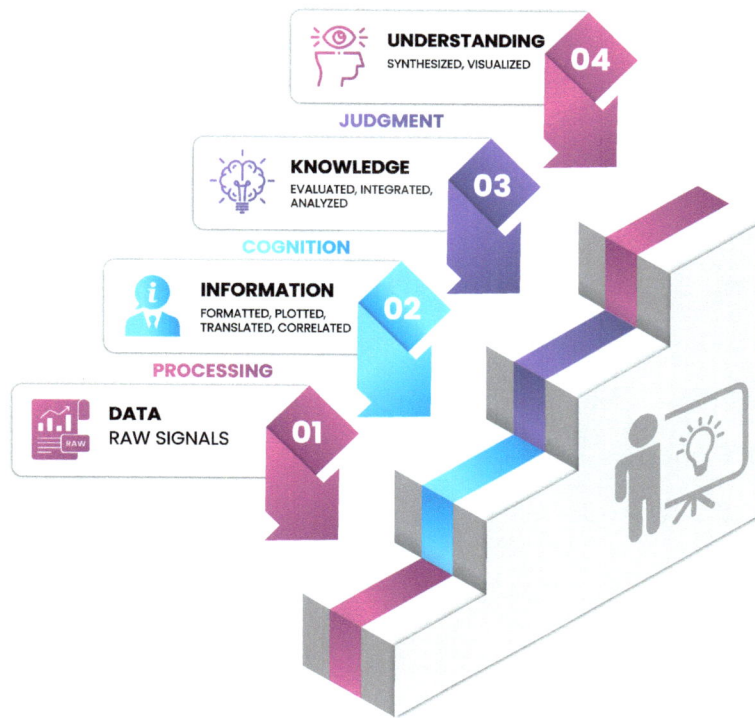

Figure 6: Information Process

The concepts of agility and velocity apply to the intelligence cycle and the capability to conduct multiple modeling and simulation iterations over a brief period, with substantial amounts of data, to render information into intelligence.

Activity-based intelligence methods integrate large volumes of multi-discipline, multi-domain data to establish and maintain a baseline that reveals patterns and anomalous behavior, identifies gaps in information to drive further collection, and provides a real-time understanding of the operational environment. Activity-based intelligence is found in the capabilities-based assessment for intelligence and can be viewed as a spiral process of intelligence fusion within the intelligence cycle for big data analysis.

Information Operations (IO) vs. Intelligence Operations

While often intertwined, information operations (IO) and intelligence operations serve distinct purposes within the cybersecurity landscape.

Information Operations (IO)

IO focuses on shaping the narrative within the information environment to achieve specific objectives. This includes both defensive measures like protecting sensitive information and offensive tactics like disinformation campaigns or psychological operations. The primary goal of IO is to influence the decision-making, perceptions, and behaviors of target audiences, whether they are adversaries, neutral parties, or even the organization's own personnel.

Intelligence Operations

Intelligence operations, on the other hand, center on the collection, analysis, and dissemination of information to gain a deeper understanding of the threat landscape. This involves identifying potential adversaries, assessing their capabilities and intentions, and proactively identifying vulnerabilities within the organization's systems and networks. The aim of intelligence operations is to provide decision-makers with actionable insights that inform risk management strategies and enable them to make informed choices to protect critical assets.

Key Differences

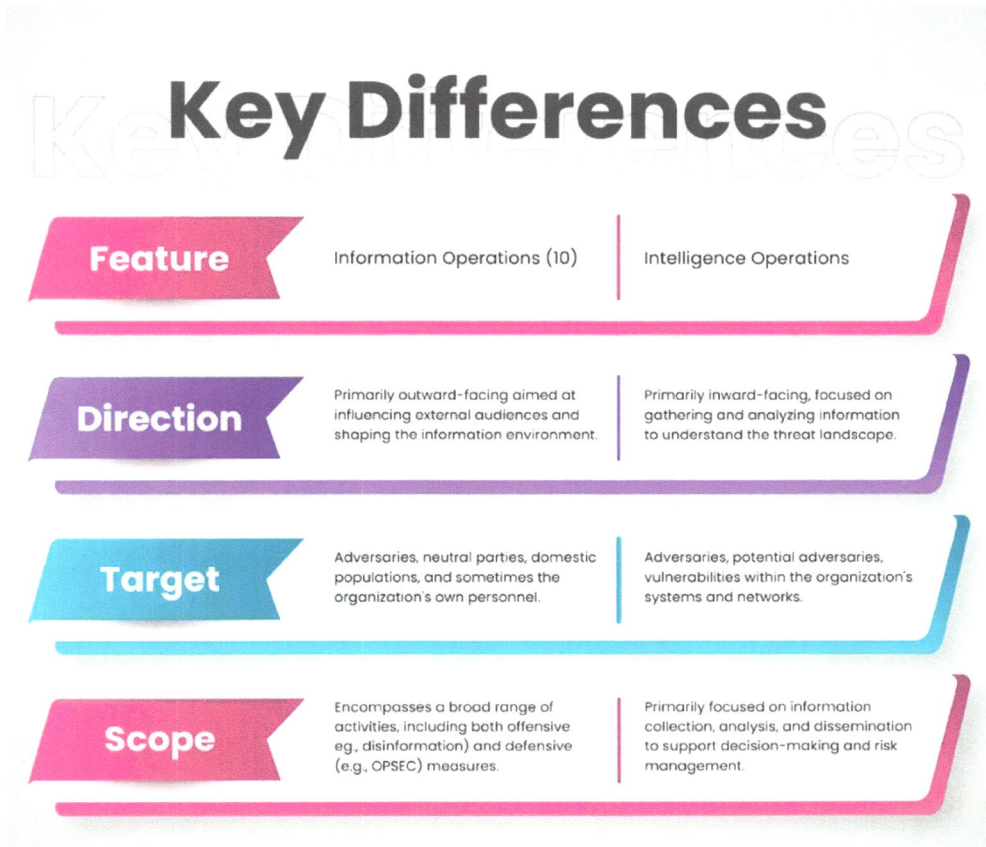

Figure 7: Key differences between IO and Intel Ops

Synergy and Interdependence

While distinct, IO and intelligence operations are highly interconnected and often work in tandem. IO strategies rely on accurate and timely intelligence to be effective, while intelligence operations can benefit from IO techniques to protect their activities and enhance their impact.

33

By understanding the nuances of IO and intelligence operations, organizations can develop a more holistic and effective approach to cyber risk management, one that leverages both proactive and reactive measures to safeguard their assets and operations.

To illustrate this crucial point, let's consider a common scenario:

Scenario: Legacy Systems in the Datacenter

Consider the Data Center with an aging yet still operational section for legacy applications. These applications might not be well-understood by current IT staff, and support documentation may also be lacking. A vulnerability is discovered in middleware common to these systems.

Risk Intelligence: Identifies the business-critical functions dependent on this legacy system. Prioritizes vulnerability assessment based on the potential business impact of an outage or breach.

Threat Intelligence: Researches if existing exploits exist in the wild, or if threat actors are actively targeting this vulnerability. This informs patching urgency.

Operational Considerations: The Data Center Operations team provides input that patching could disrupt other systems or require planned downtime. This helps balance security and operational continuity.

Decision-Making: Leaders, informed by the combined insights, can make a risk-informed decision: immediate but disruptive patching, temporary compensating controls, or a staged patching plan.

Intelligence Functions

The primary function of intelligence in cyber risk management is to support the organization's overall objectives. This includes developing a comprehensive understanding of the situation, providing early warnings of cyber risks, issues, and threats, supporting force protection measures, informing contracting and procurement decisions, and assisting in site and cyber assessments. These functions collectively contribute to a more proactive and resilient cybersecurity posture.

1. Support organization's estimate.
2. Develop the situation.
3. Provide indications and warning of Cyber Risk, Issues, and Threats.
4. Support force protection.
5. Support contracting and procurement decisions.
6. Support site and cyber assessments.

In this sequence, intelligence needs are identified; a plan is developed to satisfy these needs; and data is collected, processed into information, and converted into intelligence through analysis and synthesis.

Figure 8: Intelligence Functions

The resulting knowledge is then provided to the decision makers as an intelligence product to assist in planning or informing decisions. Automation is essential to enabling the necessary agility and velocity. Analysts are key in defining the information requirements, detection automation, and systems enrichment that ensures that the Critical Information Requirements are kept in focus and adjusting for system and personal bias.

Intelligence information is derived from intelligence data, sensor data, and field data:

- Intelligence data is derived from assets primarily dedicated to intelligence collection (e.g., imagery, electronic intercept, human intelligence [HUMINT] sources).
- Sensor data is derived from crewed and uncrewed systems used for reconnaissance, surveillance, or target acquisition (e.g., air surveillance radar, counterbattery radar, uncrewed aircraft systems [UASs], remote ground sensors).
- Field data is derived from reporting by subordinate, adjacent, or other partnered organizations.

This analysis not only drives intelligence collection and ongoing operations, but it also provides context. This allows decision makers to make rapid decisions when the indicators of Cyber Risk or compromise have been observed.

The analyst must maintain a clear picture of the contextual parameters and their own biases to ensure that the indicators are in fact true indicators and not signals mistaken for facts. This clear picture is required to produce a reliable and actionable intelligence estimate by predicting probable future environmental conditions or adversary actions and conveys these results to the commander. Because we usually understand situations best as images, intelligence should be produced and disseminated in graphic form whenever possible.

Planning and Direction

The initial step identifies intelligence needs and develops a plan for satisfying those needs.

The initial planning phase is important to get a clear understanding of the requirements for the CEO/Business. Understanding what the needs and requirements are from the beginning will help guide the intelligence process and the outcome of the needs.

Collection

Collection includes developing the required intelligence structure and identifying current and native sources of Information used to generate intelligence.

During the collection phase of the intelligence cycle the Analyst and the Cyber Security/ Architect will need to work together to ensure there is a cyber collection plan put in place to help mitigate potential threats in the cyber domain.

Processing and Exploitation

This step converts collected information into an understandable form suitable to produce intelligence. Processing is accomplished during collection or production. Data collected in a form suitable for analysis is processed automatically during collection. Other types of data require extensive processing, which can affect the timeliness and accuracy of the resulting information. Because processing and production are often accomplished by the same organization, production management generally encompasses processing functions that are required to convert raw data into a usable format.

Examples of processing and exploitation include:
- Film processing.
- Document translation.
- Signals intercept.
- Electromagnetic spectrum
- Information operations

Production Steps

Production converts data into intelligence and creates the knowledge needed for the planning and execution of operations. Intelligence must deliver knowledge, in context, in time, and in a form usable in the decision-making process. In any situation, providing timely, accurate, and relevant intelligence to and planners is a critical consideration.

Production (analysis and synthesis) can be complex, such as comprehensive and detailed Lessor Cyber Assessments required to support the planning of Data Centers in new regions, or simple, such as direct answers to rapidly changing questions needed to support the implementation of a new Fire suppression system technology.

For this reason, production is distinguished as:

- Deliberate production makes full use of available information to provide a complete and extensive product that satisfies non-time sensitive intelligence requirements (IRs). This type of production normally supports operational planning.

- Immediate production identifies information directly applicable to current operations, the information is subjected to a compressed version of production process, and the resulting product is rapidly disseminated to those affected.

The Cyber Business of Data Centers:

A Data Center is the perfect example of the interconnectivity of our digital and physical world.

A Data Center has a specific physical nature:

1. Hardware is the physical equipment that comprises the components that enable digital processes.
2. Physical security measures must protect the hardware as it is the literal embodiment of the digital world. A physical breach has the same ultimate implications as a virtual breach.
3. Temperature and power systems physically support the continuous operations of digital systems. A humidity or water failure can have the same devastating effects on a resource as a distributed denial of service attack. Both mean that the customer or the business cannot access the resources.

The Data Center has layers.

Physical threats and Digital threats can disrupt a Data Center. Fire, flood, power outages, or unauthorized access to the physical Data Center facility pose critical risks to the digital operations, data, and services housed within. While Malware, ransomware, or hacking attempts can target the digital systems within the Data Center, but their impact can manifest physically: systems shutting down, data being exfiltrated, or services becoming unavailable.

Disaster Recovery requires both digital and physical considerations.

Backups

Maintaining physical backups of data at an offsite location is a critical component of effective disaster recovery planning. These physical backup media, such as tapes or hard drives, serve as a crucial digital lifeline that can be used to restore systems and data in the event of a disaster (Veeam, 2022). By storing backups off-site, organizations can ensure that their data and systems can be recovered even if the primary data center is damaged or inaccessible.

Disaster Recovery Processes

Comprehensive disaster recovery plans must address both the physical and digital aspects of system restoration (NIST, 2016). On the physical side, this includes tasks such as restoring power, repairing, or replacing damaged hardware, and relocating personnel to an alternate site. Closely intertwined with these physical steps are the digital processes of data recovery, system reconfiguration, and service restoration. Effective disaster recovery planning requires carefully coordinating these interdependent physical and digital elements to ensure a smooth and timely recovery (Gartner, 2021).

Introducing the Data Center Model: A Framework for Risk Assessment and Mitigation

The digital era has transformed the modern business landscape, ushering in unprecedented opportunities but also exposing organizations to a growing array of cyber threats. At the heart of this digital transformation lie data centers – the physical and digital fortresses that house the critical infrastructure powering our connected world. These concentrated hubs of valuable assets and interconnected systems have become prime targets for adversaries seeking to disrupt operations, steal data, or cause financial and reputational damage.

To safeguard these critical assets, data center operators must adopt a comprehensive approach to security, one that addresses both physical and digital vulnerabilities. The data center model provides a robust framework for achieving this goal. This model offers a structured methodology for assessing and mitigating the diverse risks that threaten data centers, ensuring the availability, integrity, and confidentiality of critical data and systems.

Applying the Data Center Model to Risk Assessment

The data center model is a holistic approach to risk assessment, recognizing that threats can originate from various sources:

1. **Physical Risks:** These encompass threats to the physical infrastructure of the data center, including fire, flood, power outages, hardware failures, and unauthorized physical access. A fire in a data center could lead to extensive damage, disrupting operations and causing significant financial losses.

2. **Digital Risks:** These threats target the digital assets and systems within the data center, such as malware, ransomware, hacking attempts, data breaches, and insider threats. A successful ransomware attack could encrypt critical data, halting operations and potentially resulting in extortion demands.

3. **Operational Risks:** These risks arise from human error, misconfigurations, software vulnerabilities, and supply chain attacks. For instance, a misconfigured firewall could inadvertently expose the data center to external threats, while a vulnerability in a third-party software component could be exploited by attackers.

By analyzing the interdependencies between these risks, organizations can gain a comprehensive view of their threat landscape. This holistic understanding is crucial for developing targeted mitigation strategies that address the unique vulnerabilities of their specific data center environment.

Mitigating Risks Using the Data Center Model

The data center model doesn't just stop at risk assessment; it also guides the implementation of effective mitigation strategies.

1. **Physical Risk Mitigation:** This involves measures like investing in redundant power supplies and backup generators, implementing robust fire suppression systems, and enforcing strict physical access controls. For example, a data center could install biometric scanners and security cameras to prevent unauthorized access and monitor for suspicious activity.

2. **Digital Risk Mitigation:** This includes deploying advanced threat detection and prevention tools, such as intrusion detection systems (IDS) and next-generation firewalls, as well as implementing regular data backups and disaster recovery procedures. Additionally, robust encryption and access controls can protect sensitive data from unauthorized access.

3. **Operational Risk Mitigation:** To mitigate operational risks, organizations should conduct regular security audits and configuration reviews, enforce strong password policies, and provide comprehensive security awareness training for employees By establishing clear incident response procedures and maintaining up-to-date software, organizations can minimize the impact of human error and software vulnerabilities.

The SANS Data Center Physical Security Checklist: Practical Implementation

The SANS Data Center Physical Security Checklist outlines key elements that Data Center operators must consider when establishing a robust physical security posture. The checklist is divided into Property and People. We have created a list of nine items to ensure that physical security safeguards are integrated into our network operating systems and applications.

For Cyber Risk Intelligence we align the Cyber Security Threat with the physical vulnerabilities:

1. **Site Location**
   - **Vulnerability:** Proximity to natural disaster risks (e.g., earthquakes, floods, wildfires) or human-caused disaster risks (e.g., proximity to airports, refineries, high-traffic areas) can expose the data center to physical damage and disruption.
   - **Threat:** Cyber-attacks targeting the data center's physical infrastructure, such as hacking building management systems or disrupting power/connectivity, can exploit these location-based vulnerabilities.

2. **Perimeter Security**
   - **Vulnerability:** Inadequate perimeter security measures, such as weak fencing, lack of surveillance, or poor lighting, can allow unauthorized physical access to the data center.
   - **Threat:** Cyber-attacks targeting the access control systems or security cameras can bypass the perimeter security, enabling physical intrusion and potential data breaches or sabotage.

3. **Building Access Control**

   - **Vulnerability:** Weaknesses in access control systems, such as ineffective authentication methods or poor visitor management, can allow unauthorized personnel to gain physical access to the data center.

   - **Threat:** Cyber-attacks targeting the access control systems, such as hacking badge readers or bypassing multi-factor authentication, can exploit these vulnerabilities to gain unauthorized physical entry.

4. **Computer Rooms**

   - **Vulnerability:** Inadequate environmental controls or fire safety measures in the computer rooms can lead to physical damage or disruption of critical IT infrastructure.

   - **Threat:** Cyber-attacks targeting the building management systems or environmental control systems can manipulate temperature, humidity, or fire suppression, causing physical harm to the data center's equipment and operations.

5. **Facilities**

   - **Vulnerability:** Lack of redundancy or inadequate maintenance of critical facilities, such as cooling systems or power infrastructure, can make the data center susceptible to physical failures and disruptions.

   - **Threat:** Cyber-attacks targeting the facility management systems or industrial control systems can disrupt the data center's critical infrastructure, leading to physical outages and service interruptions.

6. **Disaster Recovery**

   - **Vulnerability:** Weaknesses in the disaster recovery plan, such as inadequate offsite backups or lack of a redundant data center, can leave the organization vulnerable to prolonged downtime and data loss in the event of a physical disaster.

   - **Threat:** Cyber-attacks targeting the disaster recovery systems or processes can compromise the organization's ability to recover from a physical disaster, exacerbating the impact and prolonging the disruption.

7. **People**

   - **Vulnerability:** Lack of security awareness, poor access controls, or insider threats among employees, service providers, or visitors can enable physical access to sensitive areas of the data center.

   - **Threat:** Cyber-attacks targeting the human element, such as social engineering or phishing, can exploit these vulnerabilities to gain unauthorized physical access or manipulate the actions of personnel.

8. **Emerging Technologies**

   - **Vulnerability:** Vulnerabilities in emerging technologies, such as AI-powered surveillance or blockchain-based access logs, can create new physical security risks if not properly implemented and secured. The adoption of IoT devices or advanced physical surveillance systems can introduce *new* cyber-physical risks alongside their potential benefits.

   - **Threat:** Cyber-attacks targeting these emerging technologies can bypass or manipulate the physical security measures, undermining the data center's overall security posture.

9. **Human Factors**

   - **Vulnerability:** Social engineering tactics, insider threats, or lack of security awareness among personnel can lead to physical security breaches or enable cyber-physical attacks.

   - **Threat:** Cyber-attacks exploiting human vulnerabilities, such as phishing or manipulation of employees, can result in physical security incidents, data breaches, or operational disruptions.

By evaluating these components, we will equip you with the knowledge and strategies necessary to understand how to fortify your resources against a wide range of cyber threats that include physical attack vectors. Through real-world case studies, industry best practices, and expert insights, you will learn how to adapt this model to the unique requirements of your organization, ensuring the continuous availability, integrity, and confidentiality of your most valuable digital assets.

As we progress through the chapters, this model will serve as the context upon which we build a comprehensive understanding of cyber risk intelligence, empowering you to proactively assess your digital fortress against the evolving threat landscape.

Cyber + Risk + Intelligence

Cyber Risk Intelligence (CRI) team is a front-line defense against digital threats. Combining diverse expertise in threat analysis, risk management, and cybersecurity, they proactively identify, assess, and mitigate the risks targeting our infrastructure. They collaborate to proactively identify, assess, and mitigate cyber threats targeting our infrastructure. This team may also include incident responders and communication specialists to ensure a comprehensive and effective approach. They leverage their collective knowledge to provide insightful, actionable intelligence that informs leadership decisions and strengthens our organization's overall cyber resilience.

The core expertise must include a persona with systems level knowledge of the systems that we would like to assess. However, where there are partners with clear and transparent communication lines, the CRI team can adjust to expand and contract as necessary to complete the assessments and round out the insights.

It is essential to think of Cyber Risk Intelligence as both a team of individuals and a functional capability. This means that the redundancy starts at the human resources. The tooling and technology are enablers rather than requirements.

No matter the implementation model, the key is to ensure that the CRI function is well-integrated into the organization's overall security strategy and that it has the resources and support it needs to be effective.

When we use the example of a Data Center, the CRI Team would require focus in construction, functionality, and layout of those Data Centers. A Data Center is a physical facility housing an organization's critical IT systems and data. A part-time or ad hoc team member could provide this knowledge. This would allow for a member of the Data Center

Operations team while working closely with cyber specialists to experience the strategic level of the organization while providing key insights and data to the team as a "Human Intelligence" signal.

Real life outcomes include identification of vulnerabilities in legacy software that can lead to prioritized patching and threat mitigation, preventing a potential data breach.

In today's landscape of scarce resources, the modern CRI team prioritizes understanding the process and effectively gathering the raw data that drives strategic insights. The modern approach requires that the human understand the process and operates within it. They are aware enough to relay or transmit raw data that can be processed by the team to create the consistent and relevant insights about the environment, the threats, and the vulnerabilities.

Core Skills and Desired Expertise

The core skills and areas of expertise that members of a Cyber Risk Intelligence team typically possess:

- **Intelligence Analysis:** Experts in collecting, analyzing, and interpreting data from various sources to identify patterns, trends, and potential threats.
- **Risk Management:** Professionals skilled in assessing cyber risks, vulnerabilities, and the potential impact of threats on the organization.
- **Cybersecurity:** Specialists in technical cybersecurity concepts, attack vectors, threat actor tactics, and defensive measures.
- **Incident Response:** (Optional) Individuals experienced in responding to cyber incidents, understanding how to contain and mitigate threats.
- **Communication:** Team members with strong communication skills to effectively share intelligence insights with leadership and technical teams.

When we design the team that fulfills this function, the following questions must be asked:

1. *Who should own the teams?*
2. *Have we eliminated blockers to collaboration?*
3. *Risk and Threat are information flow dependent, what tools are enabling data sharing?*
4. *Is there both coverage and expertise overlap?*

Mission, Methodology, Their Drive, For Implementers

Mission:

The Cyber Risk Intelligence team's primary mission is to provide seamless, actionable, and timely intelligence to leadership at every level, enabling informed decision-making that protects the organization's critical assets, infrastructure, and people.

How They Achieve This:

Leveraging Internal Knowledge: The team collects, processes, analyzes, and interprets information from both internal and external sources, drawing upon its deep understanding of the organization's unique systems, processes, and vulnerabilities.

Objectives:

- To protect life, critical infrastructure, facilities, and resources.
- To reduce uncertainty and enhance organizational resilience.
- To provide actionable intelligence that enables proactive threat mitigation and strengthens the organization's security posture.

Methodology:

The Cyber Risk Intelligence (CRI) team functions as a specialized intelligence cell focused on risk assessment. Their primary objective is to develop a comprehensive and accurate understanding of the business and its environment to accurately evaluate the likelihood and potential impact of identified risks.

The CRI team provides essential insights to the business, while also collaborating with the Threat Intelligence (TI) team. TI leverages CRI's analysis to implement counterintelligence measures and continuously improve processes and monitoring. TI is primarily concerned with identifying emerging threats, understanding threat actors, and directly protecting the organization from them. Conversely, the Business Intelligence team focuses on analyzing the market, customers, and competition to optimize operations.

Integrated Approach

The Cyber Risk Intelligence team's methodology combines the best practices of intelligence, risk management, and cybersecurity for a comprehensive approach. This, coupled with their in-depth knowledge of the organization's specific cyber landscape, enables targeted threat identification and assessment. This multidisciplinary approach is essential. It ensures the methodology is robust enough to tackle diverse risks with the best tools from each field.

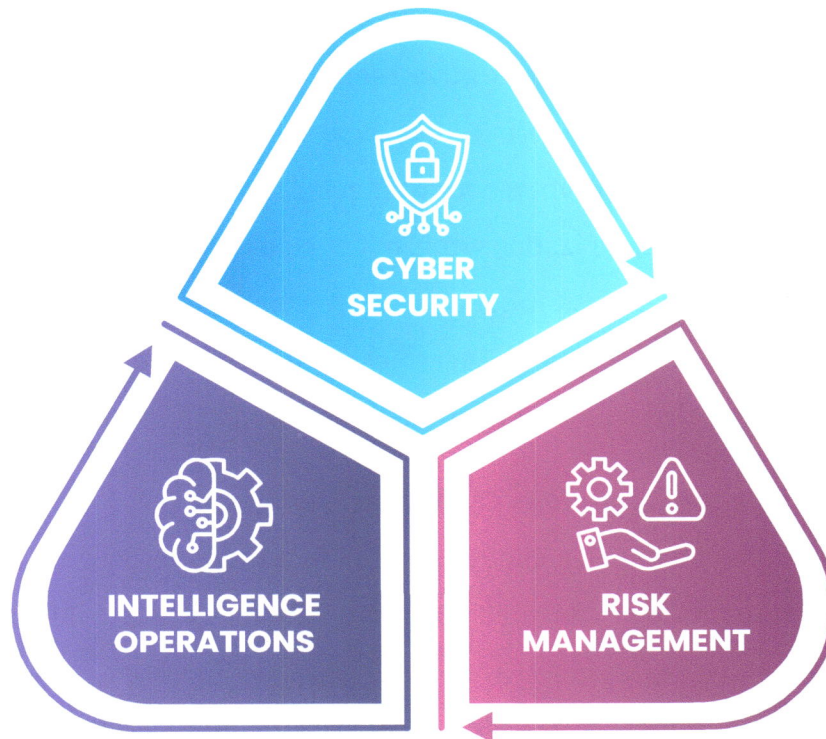

Figure 9: The Triad of Cyber Resilience

Their Drive:

The team's deep knowledge of the organization's specific cyber landscape enables targeted threat identification and assessment. The Cyber Risk Intelligence team understands the critical importance of protecting the organization's valuable assets. The ever-evolving threat landscape fuels their proactive approach to identifying and mitigating cyber risks through informed, data-driven decisions. Deeply integrated within the organization, they collaborate effectively with all stakeholders to ensure the highest levels of protection. The team's systematic analysis, combined with their operational agility and zero-tolerance for risk, provides leadership with the insights needed to safeguard the business and its customers.

For Implementers:

People: Building a Skilled and Collaborative Team

The heart of CRI lies in a team of experts with diverse skill sets. Intelligence analysts, risk management professionals, and cybersecurity specialists collaborate to identify, assess, and mitigate risks. In larger organizations, this might be a dedicated, full-time team. In smaller settings, it could be a cross-functional effort where individuals contribute their expertise alongside their primary roles. Regardless of the model, open communication, shared knowledge, and a commitment to continuous learning are essential for success.

Process: The Intelligence Operations Cycle as a Framework

CRI is not just about the people; it's about establishing a robust process that seamlessly integrates into your operations. The Intelligence Operations Cycle provides a proven framework for this:

1. **Planning & Direction:** Clearly define objectives, identify stakeholders, and establish metrics for success.

2. **Collection:** Gather relevant data from various sources, including threat intelligence feeds, internal logs, and external partners.

3. **Processing & Exploitation:** Analyze the collected data to identify patterns, anomalies, and potential threats.

4. **Analysis & Production:** Translate the processed data into actionable intelligence that informs decision-making.

5. **Dissemination:** Share the intelligence with relevant stakeholders in a timely and understandable manner.

This cyclical process enables continuous improvement, ensuring that your CRI capabilities adapt to the ever-changing threat landscape.

Technology: Empowering Your Team

While people and processes are essential, technology is a force multiplier in CRI. Leverage tools like:

- **Threat Intelligence Platforms:** Aggregate and analyze threat data to identify emerging risks.

- **SIEM Systems:** Correlate security events to detect anomalies and potential incidents.

- **AI and Machine Learning:** Automate threat detection and analysis, uncovering patterns that humans might miss.

These technologies empower your team to efficiently process vast amounts of data, detect threats earlier, and make informed decisions.

No One-Size-Fits-All:

The beauty of CRI is its adaptability. Whether you have a dedicated team, rely on cross-functional collaboration, or adopt a hybrid approach, the key is to tailor the implementation to your organization's specific needs and resources. The ultimate goal is to create a cyber risk intelligence function that is well-integrated into your overall security strategy and has the support it needs to thrive.

By focusing on the people, processes, and technology of CRI, you can build a robust defense against cyber threats and safeguard your organization's critical assets.

Intelligent Cyber Defense: A Data-Driven Approach to Risk Management

What is Intelligent Cyber Defense?

Intelligent cyber defense represents the latest evolution in cybersecurity, leveraging cutting-edge technologies to enhance traditional security measures. While artificial intelligence (AI) and machine learning (ML) are undoubtedly significant components of this evolution, they are not the sole driving force.

Intelligent cyber defense builds upon decades of progress in information security, incorporating advancements in data analytics, automation, and threat intelligence. It represents a shift from reactive measures to a proactive, data-driven approach that empowers organizations to anticipate and mitigate risks before they can be exploited.

By analyzing vast amounts of data, intelligent systems can detect anomalies, identify patterns, and even predict potential threats, allowing organizations to respond more quickly and effectively. This approach complements human expertise, enabling security teams to focus on higher-order tasks while automation handles routine threat detection and analysis.

Importantly, intelligent cyber defense is not about engaging in offensive cyber warfare or counterattacks. It focuses on using the full spectrum of tools and techniques available—including AI and ML—to gain a deeper understanding of the threat landscape, identify vulnerabilities, and proactively mitigate risks before they can be exploited.

The Role of Operational Risk Management

Operational risk management (ORM) plays a crucial role in intelligent cyber defense. By systematically identifying, assessing, and controlling risks across an organization's operations – including technology, processes, and people – ORM ensures that security measures are aligned with business goals and priorities. Integrating intelligent cyber defense with ORM creates a powerful synergy. Intelligent defense provides the insights and tools needed to identify and prioritize threats, while ORM translates these insights into actionable strategies that protect the organization's critical assets and operations.

Key Terms

- **Course of Action (COA):** Possible actions an adversary might take.

- **Critical Information Requirements (IRs):** Key questions leaders need to be answered to make informed decisions.

- **Intelligence:** Data analyzed and placed in context to provide insights for decision-making.

The Synergy of Intelligent Defense and Operational Risk Management

The combination of intelligent cyber defense and operational risk management creates a powerful synergy. Intelligent defense provides the insights and tools needed to identify and prioritize threats, while ORM ensures that these insights are translated into actionable strategies that protect the organization's critical assets and operations.

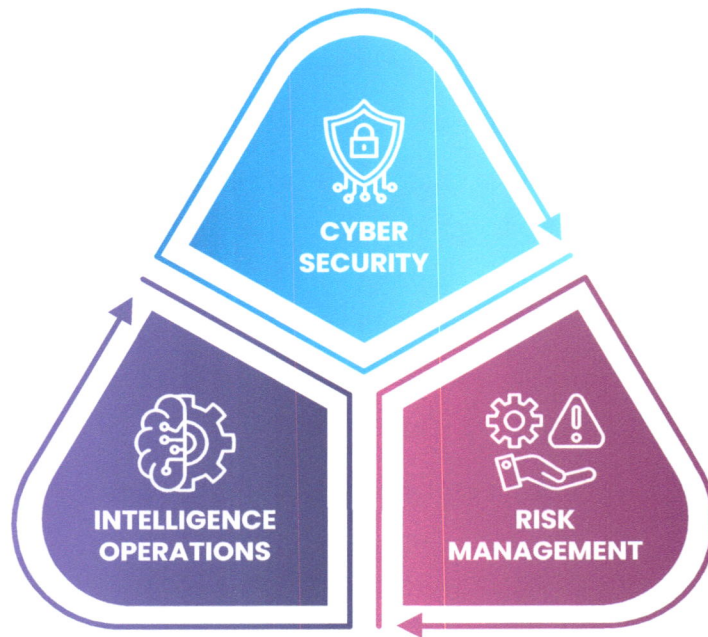

Figure 10: The Synergy of Intelligent Cyber Defense
and Operational Risk Management

Key Terms

- **Course of Action (COA):** Possible actions an adversary might take.
- **Critical Information Requirements (IRs):** Key questions leaders need answered to make informed decisions.
- **Intelligence:** Data analyzed and placed in context to provide insights for decision-making.

Operations

In the realm of both cyber and cyber risk intelligence, the ability to gather and interpret information about the environment, adversaries, and the business is paramount. As the Marine Corps Warfighting Publication 2-10 aptly states, *"The fog and friction of war will never allow a perfect picture of the operational environment."* This principle applies equally to the cyber domain, where the constantly evolving threat landscape creates uncertainty and challenges the ability to fully understand the situation.

Effective cyber risk intelligence (CRI) operates with this uncertainty in mind, focusing on the most critical information needed for the organization's decision-making. This involves:

- **Collecting Essential Information:** Targeted collection aligned with leadership's critical information requirements (IRs).
- **Contextualizing Data:** Analyzing information within the broader operational and cyber landscape to uncover hidden meanings and connections.
- **Communicating Intelligence:** Providing clear, timely, and relevant reports that directly support decision-making.

This approach ensures that resources are focused on gathering and analyzing the most relevant information, enabling organizations to make informed decisions despite the inherent uncertainties of the cyber environment.

<u>Intelligence has two basic complementary objectives.</u>

Intelligence in the cyber realm has two fundamental, complementary objectives:

1. **Primary Objective: Cybersecurity Intelligence**
   - **Focus:** Delivering accurate, timely, and relevant information about cyber threats (adversaries, malware, vulnerabilities) and the broader technological landscape.

- **Goal:** Support decision-making by reducing uncertainty about the organization's cybersecurity posture, enabling leaders to prioritize risks and allocate resources effectively.
- **Outcome:** This directly enhances the organization's ability to detect, prevent, and respond to cyberattacks, safeguarding critical data and systems.

2. **Secondary Objective: Security Intelligence (with emphasis on Supply Chain)**
   - **Focus:** Protecting the organization by proactively identifying and mitigating risks within the supply chain, including vulnerabilities in third-party vendors, software, and hardware.
   - **Goal:** Safeguard the organization from external threats by building resilience into the supply chain, ensuring the integrity of products and services, and minimizing disruption.
   - **Outcome:** This proactive approach helps prevent supply chain attacks, which can be devastating due to their potential to compromise multiple organizations simultaneously.

Both objectives are essential for a comprehensive cybersecurity strategy. Cybersecurity intelligence focuses on the immediate threats and vulnerabilities an organization faces, while security intelligence takes a broader view, considering the risks that may arise from external dependencies and partnerships. Together, they provide a holistic understanding of the threat landscape and enable organizations to make informed decisions that protect their assets, operations, and reputation.

Counterintelligence is information gathering and activities conducted to identify, deceive, exploit, disrupt, or protect against espionage, other intelligence activities, or sabotage conducted for or on behalf of adversaries or threats.

Counterintelligence supports the protection of allied systems by identifying vulnerabilities and assisting in developing appropriate plans to enhance a unit's posture against these threats. Counterintelligence (CI) activities consist of active and passive measures, functions, and services that support the four missions of CI:

1. Countering espionage, international terrorism, and the CI insider threat
2. Providing support to force protection
3. Providing support to defense critical infrastructure.
4. Providing support to research, development, and acquisition to—

    o Deny adversaries and enemies information they can use to increase the effectiveness of hostile operations against friendly forces.

    o Detect and neutralize adversary intelligence collection.

    o Deceive the adversary as to friendly capabilities and intentions.

Counterintelligence, in conjunction with operations support and planning, can magnify the effects of uncertainty on the adversary's decision-making process, thereby contributing to the success of friendly operations.

Intelligence Agility in the Cyber Domain

The cyber threat landscape is ever evolving. Effective cyber intelligence operations must be adaptable to counter new tactics and techniques. Real-time analysis and continuous refinement of intelligence processes are critical for staying ahead of attackers.

Benefits of Integrating Intelligence and Operational Risk

This scenario highlights how cybersecurity extends beyond technology alone. Here is how this integrated approach improves outcomes:

- **Reduced Uncertainty:** Cyber Risk Intelligence reduces uncertainty by quantifying threat likelihood and business impact. Informed decisions are easier to make.
- **Proactive Approach:** By understanding the threat landscape and business context, security efforts shift from reactive to proactive risk mitigation.
- **Efficiency:** The CRI team prioritizes work based on potential impact, making the most of limited resources in a constantly shifting threat environment.

Key Takeaways

Modern cyber defense is not just technical. An integrated approach ensures defenses align with the unique business context and threat landscape.

Collaboration between Risk Intelligence, Threat Intelligence, and operational teams is vital for maximizing protection and efficiency.

Part I

Essentials of Cyber and Physical Risk Monitoring

Introduction to Cybersecurity Risks

In a single day, a ransomware attack can bring a global corporation to its knees. The digital era is not just about exploiting opportunities; it is a battleground where cyber risks constantly threaten your organization. From phishing scams to stealthy state-sponsored attacks, the dangers are ever-real – and growing. In this environment, proactive cyber risk monitoring isn't a luxury, it's a necessity. The exponential rise in cybercrime, projected to cost the world over $10 trillion annually by 2025, underscores a critical era where digital defense mechanisms become paramount.

Figure 11: Global Cyber Threat Landscape: A Visual Overview

Cyber threats, ranging from phishing and ransomware to insider threats and advanced persistent threats (APTs), have not only become more frequent but have also surged in sophistication, targeting a wide spectrum of organizations, regardless of their size or sector.

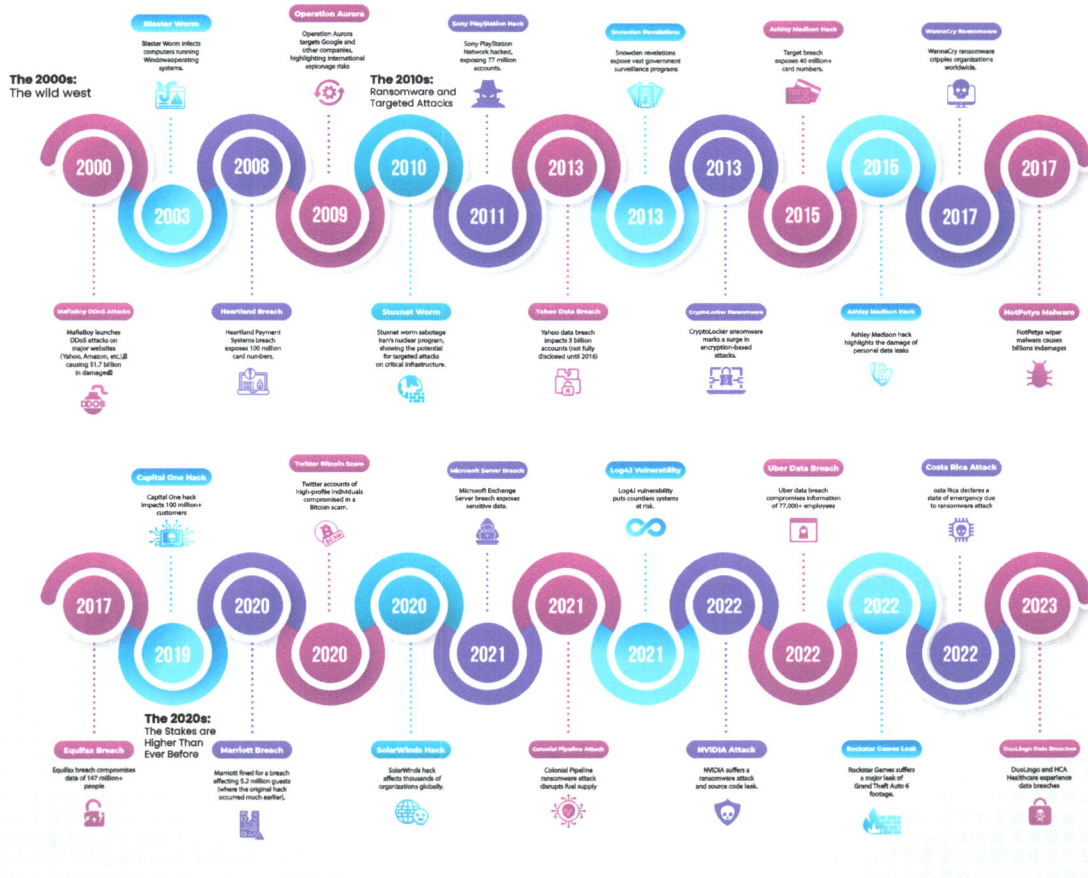

Figure 12: Evolution of Cyber Attacks: Increasing Sophistication and Complexity

Even the motivation and characterization of the threat activities have matured. We are now concerned about Acts of Cyberaggression.[1]

The evolving threat landscape is characterized by a diverse set of cyber risks that pose significant financial, reputational, and operational damages to organizations.

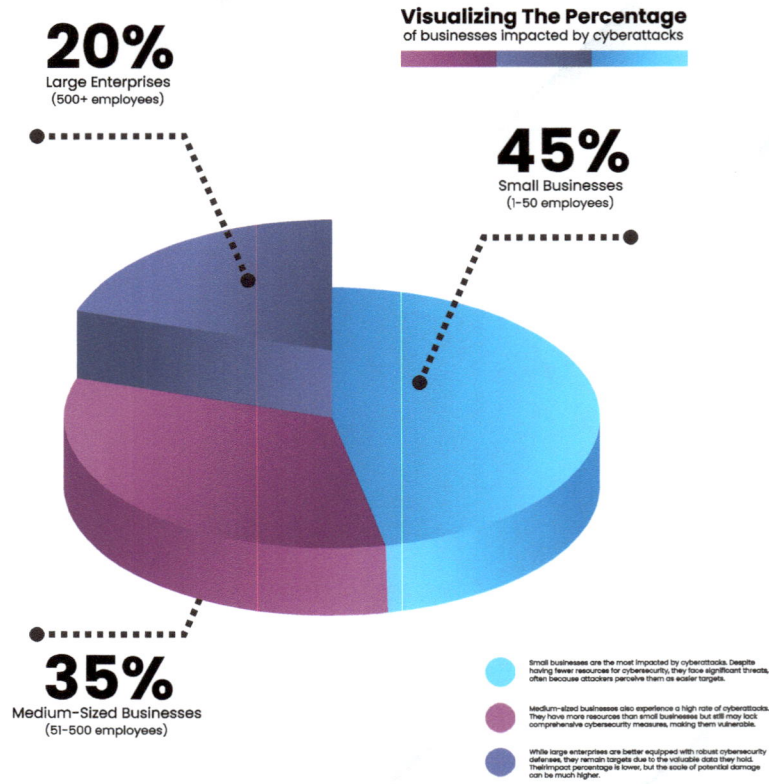

Visualizing The Percentage
of businesses impacted by cyberattacks

20%
Large Enterprises
(500+ employees)

45%
Small Businesses
(1-50 employees)

35%
Medium-Sized Businesses
(51-500 employees)

Small businesses are the most impacted by cyberattacks. Despite having fewer resources for cybersecurity, they face significant threats, often because attackers perceive them as easier targets.

Medium-sized businesses also experience a high rate of cyberattacks. They have more resources than small businesses but still may lack comprehensive cybersecurity measures, making them vulnerable.

While large enterprises are better equipped with robust cybersecurity defenses, they remain targets due to the valuable data they hold. The impact percentage is lower, but the scale of potential damage can be much higher.

Figure 13: The Multifaceted Impact of Cyberattacks on Organizations

Notable incidents, such as the SolarWinds cyberattack, the Equifax data breach, and the WannaCry ransomware attack, exemplify the dire consequences of inadequate cyber risk management. These examples highlight the critical need for organizations to address cyber risks promptly and effectively to safeguard their assets, reputation, and growth. This is not limited to protection of the digital assets, but also the physical assets that enable the digital. Physical threats are as relevant to Cyber Defense and Offense as digital threats.

Against this backdrop, the development and implementation of a comprehensive Cyber Risk Monitoring Plan (CRMP) emerges as indispensable strategies for contemporary organizations. A CRMP serves not only to identify and analyze potential cyber risks but also to implement proactive measures that enhance an organization's cybersecurity posture. The primary purpose of such a plan is to establish a structured and systematic approach to managing cyber risks, ensuring that organizations can maintain stability and protect critical operations in the face of ever-evolving cyber threats.

By following a meticulous methodology and adhering to the outlined guidelines and strategies, organizations are empowered to develop a CRMP tailored to their unique needs. This plan not only identifies, analyzes, and monitors cyber risks but also presents these risks to leadership for appropriate action, thereby safeguarding the organization's reputation, assets, and growth. The importance of a CRMP cannot be overstated, as it encompasses the core principles and objectives critical to enhancing an organization's cybersecurity defenses and resilience.

As we delve deeper into the development, implementation, and maintenance of an effective CRMP, it becomes clear that understanding the foundational concepts of cybersecurity risks is the first step towards a more secure digital future. The subsequent chapters of this book will explore the various facets of a CRMP, including risk identification, analysis, mitigation strategies, and the significance of an adaptive, continuous improvement ethos. Through a comprehensive approach, this book aims to equip organizations with the knowledge and tools necessary to anticipate, prevent, and respond to cyber risks proactively, fostering a more resilient digital landscape.

Scenario: Phishing Attack and Incident Response

Threat: Phishing attacks are fraudulent emails designed to trick users into revealing sensitive information or clicking malicious links.

Potential Impact: These attacks can lead to data breaches, ransomware infections, and financial losses. The rising costs of data breaches, as depicted in Figure 14, highlight the financial impact of cyberattacks on organizations.

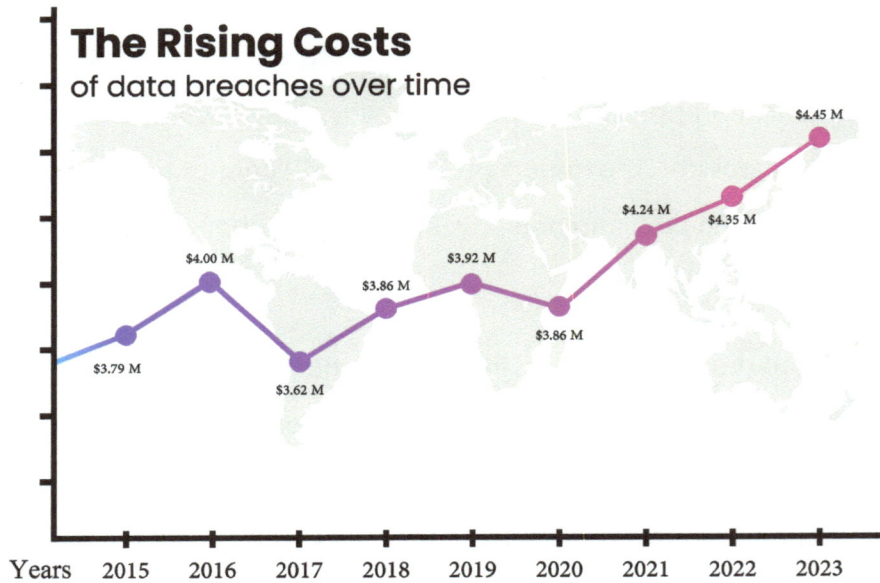

The Rising Costs
of data breaches over time

- 2015: $3.79 M
- 2016: $4.00 M
- 2017: $3.62 M
- 2018: $3.86 M
- 2019: $3.92 M
- 2020: $3.86 M
- 2021: $4.24 M
- 2022: $4.35 M
- 2023: $4.45 M

Years

Figure 14: Rising Costs of Data Breaches

Scenario: A small manufacturing company receives a seemingly legitimate email from a supplier, requesting an urgent update to their bank account details. The company's IT manager, Jeff, notices some unusual network activity around the same time.

Consequences: Without a well-defined incident response plan or clear understanding of typical vendor communication channels, Jeff is forced to make a snap decision. He shuts down key systems to isolate the potential breach, grinding production to a halt. However, further investigation reveals the email was a phishing attempt, and the network activity was unrelated. The company loses valuable production time due to the overly cautious shutdown, highlighting the need for a more nuanced response informed by risk monitoring data.

Cyber Risk Focus

To effectively manage cyber risks, it is crucial for organizations to identify and focus on the types of threats and vulnerabilities that pose the greatest risk to their operations. This chapter provides an overview of common types of cyber risks, highlighting their potential impact on businesses and emphasizing the importance of addressing these risks in a timely and systematic manner.

Understanding Cyber Risk

As an organization, your digital assets are not only your resources, but the lifeblood of your operation. Cyber risks encompass a broad spectrum of threats, vulnerabilities, and potential impacts that necessitate vigilant monitoring and strategic management.

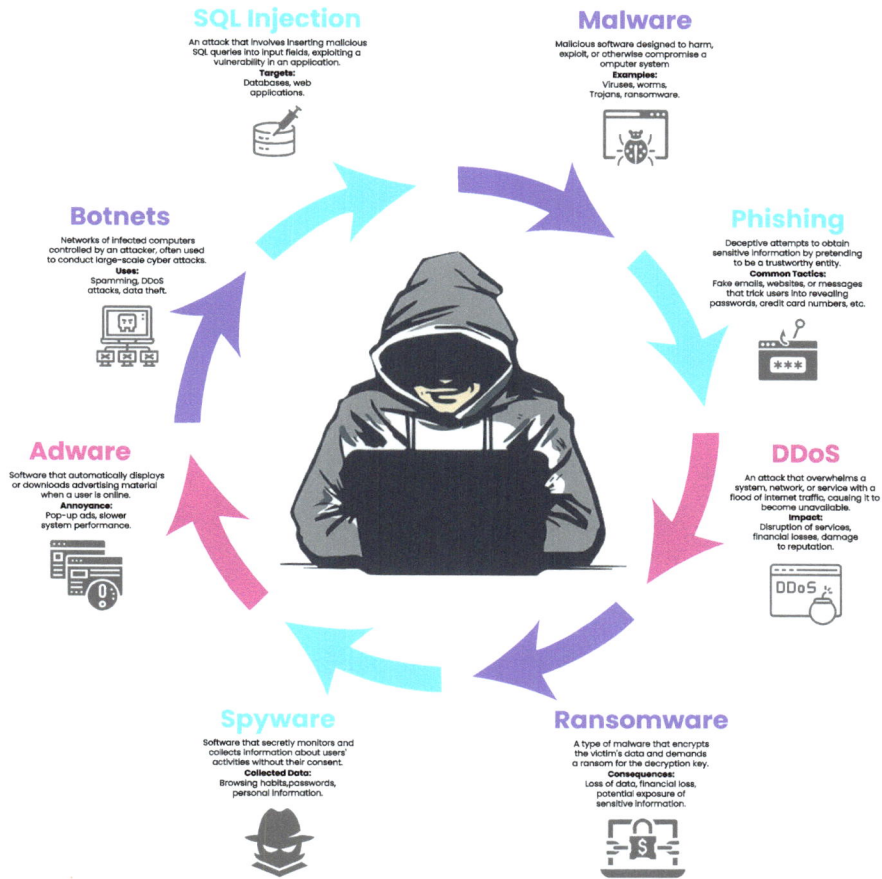

Figure 15 : Cyber Threat Landscape: A Typology of Attacks

As Figure 15 illustrates, phishing and social engineering attacks are among the most prevalent cyber threats, often serving as entry points for more sophisticated attacks.

Phishing and Social Engineering Attacks

Phishing attacks can cost companies millions. These attacks do not simply target digital entry points but also human behavior for the purposes of gaining physical access. Among the most pervasive threats are phishing and social engineering attacks, which exploit human psychology to gain unauthorized access to sensitive data. These attacks not only jeopardize financial assets but also erode the trust and integrity of an organization. Proactive monitoring for such threats enables organizations to swiftly mitigate potential breaches, safeguarding both data and reputation.

Definition: Phishing involves fraudulent emails or websites designed to trick individuals into revealing personal information or login credentials. Social engineering attacks manipulate individuals into performing actions or divulging confidential information.

Example: Colonial Pipeline Ransomware Attack (2021) – Attackers gained access to the company's systems through a compromised VPN account, likely due to a phishing attack. This led to a major supply disruption. [Include citation]

Ransomware Attacks

Ransomware attacks can cripple a business.

Ransomware attacks represent a significant threat, disrupting business operations and demanding substantial financial ransoms for the retrieval of encrypted data. The ability to quickly identify and counteract ransomware threats is crucial in minimizing operational downtime and protecting sensitive information, thus maintaining business continuity.

Definition: Ransomware is malicious software that encrypts files or systems, holding them hostage until a ransom is paid.

Example: JBS Meatpacking Attack (2021) – A ransomware attack forced the shutdown of the global meat supplier, impacting operations in the US, Australia, and Canada.

Insider Threats

Insider threat can undermine a company reputation.

Insider threats, stemming from within the organization, pose a unique challenge, potentially leading to the unauthorized disclosure of critical information and intellectual property theft. Early detection of suspicious insider activities allows for the timely implementation of countermeasures, securing valuable assets and ensuring operational integrity.

Definition: Insider threats refer to security risks from individuals within an organization, such as employees, contractors, or trusted partners.

Example: The Edward Snowden Leaks (2013) – A former NSA contractor leaked classified information, revealing extensive government surveillance programs.

Advanced Persistent Threats (APTs)

Advanced Persistent Threats (APTs) are characterized by their stealth and persistence, aiming to infiltrate systems over extended periods. Effective monitoring of APTs is vital for organizations to detect, respond to, and mitigate these sophisticated threats before they inflict long-term damage.

Definition: APTs are highly targeted cyberattacks, often carried out by nation-states or organized crime groups, which aim to remain undetected while stealing data or disrupting systems over a prolonged period.

Example: Operation Aurora (2009-2010) – A series of APTs targeting Google, Adobe, and other major companies believed to be orchestrated by the Chinese government.

Third-Party and Supply Chain Risks

Your business is not an island!

Vulnerabilities in partners' systems can leave your data exposed. Even the companies you depend on might be unknowingly leaving a backdoor open for attackers. Vigilant monitoring and management of these risks are essential in maintaining control over data and operations, thus mitigating potential breaches and disruptions. **However, cyber risks aren't solely confined to the digital realm. Physical threats to the infrastructure that houses your data and systems are equally critical to consider. An unauthorized person gaining access to a server room can be just as devastating as a skilled hacker remotely infiltrating your network.** This underscores the importance of the next topic.

Definition: Third-party risks are security vulnerabilities introduced by external partners, suppliers, or vendors who have access to your systems or data. Supply chain risks refer to broader disruptions caused by compromised links in your supply chain.

Example: SolarWinds Attack (2020) – A supply chain attack where hackers compromised SolarWinds' software updates, affecting thousands of organizations.

Types of Cyber Risks and Impact

In this section, we discuss some of the impact of the most prevalent cyber risks that organizations should be aware of and prepared to address:

1. Phishing and Social Engineering Attacks

   - **Business Impact:** Phishing and social engineering attacks can lead to unauthorized access to sensitive information, financial losses, and damage to an organization's reputation.

   - **Business Benefit of Monitoring:** By monitoring for and addressing these threats, organizations can reduce the risk of successful attacks and minimize potential consequences, such as data breaches and financial losses.

2. Ransomware Attacks

   - **Business Impact:** Ransomware attacks can disrupt business operations, lead to financial losses, and damage an organization's reputation.

   - **Business Benefit of Monitoring:** Timely identification and mitigation of ransomware threats can minimize downtime, protect sensitive information, and maintain business continuity.

3. Insider Threats

   - **Business Impact:** Insider threats can result in unauthorized access to sensitive information, intellectual property theft, and operational disruptions.

   - **Business Benefit of Monitoring:** Monitoring for insider threats allows organizations to detect suspicious activities early, reducing the potential impact and helping to safeguard valuable assets and information.

4. Advanced Persistent Threats (APTs)

   - **Business Impact:** APTs can lead to long-term security breaches, data theft, and operational disruptions.

- **Business Benefit of Monitoring:** By monitoring for APTs, organizations can detect and respond to these stealthy threats more effectively, reducing the likelihood of successful attacks and minimizing potential consequences.

5. Third-Party and Supply Chain Risks

- **Business Impact:** Third-party and supply chain risks can result in unauthorized access to sensitive information, disruptions to operations, and damage to an organization's reputation.

- **Business Benefit of Monitoring:** Monitoring and managing third-party risks helps organizations maintain control over their data and operations, reducing the likelihood of breaches and disruptions due to third-party vulnerabilities.

Importance of Addressing Cyber Risks

In an increasingly connected world, organizations must be proactive in identifying and addressing cyber risks to protect their assets and operations. Failing to do so can result in significant consequences, as evidenced by these high-profile examples:

1. **Financial Losses:** Cyber-attacks can lead to direct financial losses through theft, extortion, or operational disruptions. For instance, the 2017 Equifax data breach resulted in costs exceeding $1.7 billion due to legal expenses, cybersecurity improvements, and damage control efforts.

2. **Reputational Damage:** Security breaches can erode customer trust and damage an organization's reputation, potentially leading to lost business and revenue. After the 2013 Target data breach, the company suffered a loss in consumer trust and a drop in sales, with its stock price falling by more than 10% in the months following the incident.

3. **Legal and Regulatory Consequences:** Organizations that fail to adequately protect sensitive information may face legal and regulatory penalties, including fines and sanctions. In 2020, British Airways was fined £20 million by the UK's Information Commissioner's Office for failing to protect customer data during a 2018 cyber-attack.

4. **Operational Disruptions:** Cyber-attacks can disrupt business operations, causing downtime and loss of productivity. The 2017 NotPetya ransomware attack on global shipping giant Maersk led to massive disruptions in the company's operations, with an estimated financial impact of $300 million.

By implementing a robust Cyber Risk Monitoring Plan, organizations can proactively identify, assess, and manage cyber risks, reducing their overall risk exposure and minimizing the potential consequences of cyber incidents.

Principles of a Cyber Risk Monitoring Plan

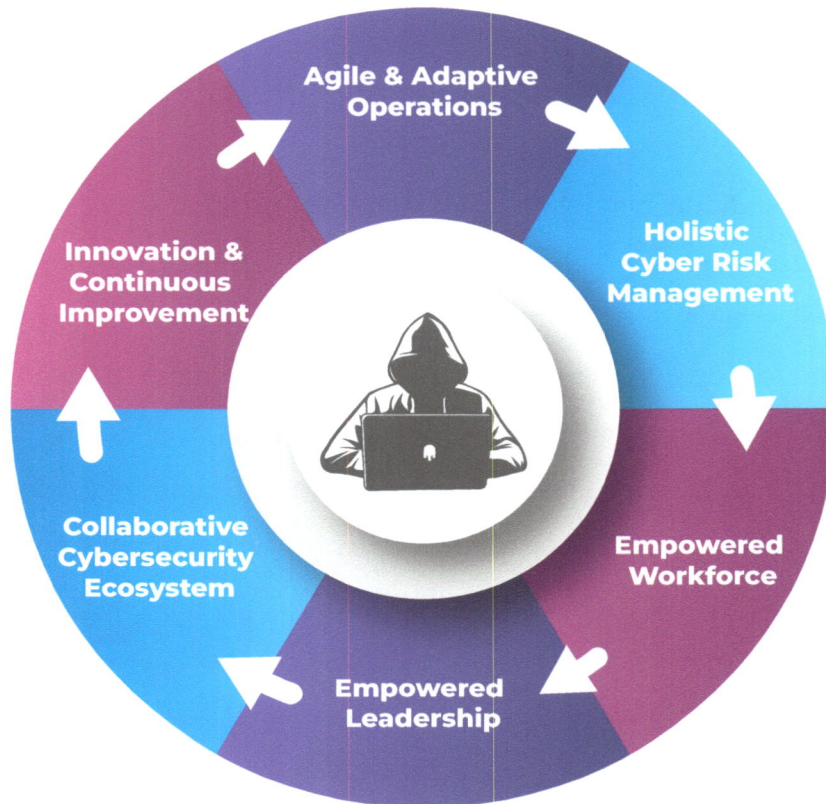

Figure 16: Principals of Cyber Risk Monitoring

The Imperative of Cyber Risk Management

The necessity of addressing cyber risks is underscored by numerous high-profile incidents, exemplified by the Equifax data breach and the WannaCry ransomware attack. These incidents illustrate the severe consequences of inadequate cyber risk management, including financial losses, reputational damage, and legal ramifications. Implementing a robust Cyber Risk Monitoring Plan enables organizations to proactively identify, assess, and manage cyber risks, thereby reducing overall exposure and minimizing potential impacts.

The digital era demands a comprehensive understanding and proactive management of cyber risks to protect organizational assets, reputation, and operations. We will delve deeper into the methodologies and strategies essential for developing, implementing, and maintaining an effective Cyber Risk Monitoring Plan. Through a combination of vigilance, strategic planning, and continuous improvement, organizations can fortify their defenses against the ever-evolving threat landscape, ensuring a more secure and resilient digital future.

Part II

Designing a Comprehensive Cyber Risk Monitoring Strategy

Scenario: Resource Allocation and Network Security

Scenario: A rapidly growing software startup is considering migrating its customer database to a cloud storage provider. The CTO, David, is impressed with the cost-effectiveness and scalability the cloud solution offers.

Threat: Security vulnerabilities in a third-party vendor's systems can create a backdoor for attackers to infiltrate your network.

Potential Impact: Supply chain cyberattacks can lead to data breaches, malware infections, and operational disruptions.

Consequences: Without thoroughly evaluating the cloud provider's security practices and compliance certifications, David might choose a seemingly attractive option that exposes the company's sensitive data to a potential breach. Cyber risk monitoring would highlight the importance of vendor security audits before critical data is entrusted to a third party.

Mitigating Third-Party Risks: To address these risks, organizations can:
- Conduct Vendor Security Audits
- Implement Strong Contractual Agreements
- Continuous Monitoring

A Cyber Risk Monitoring Plan (CRMP) is a strategic tool for identifying, assessing, and mitigating these risks before they cause damage to your organization.

Introduction to Cyber Risk Monitoring

The ever-evolving digital landscape presents organizations worldwide with a constant stream of cyber threats. High-profile events like the SolarWinds cyberattack underscore the urgent need for robust cyber risk monitoring and management. These attacks exposed the vulnerability of digital infrastructures and the devastating consequences of neglecting cybersecurity.

Purpose of the Cyber Risk Monitoring Plan

The primary purpose of a Cyber Risk Monitoring Plan is to provide organizations with a structured and proactive approach to identifying, assessing, and managing cyber risks. The steps below enable organizations to:

1. Understand their unique cybersecurity risk landscape.

   - **Business Benefit:** This understanding enables organizations to make informed decisions about their cybersecurity investments and strategies, ensuring resources are allocated where they can have the greatest impact.

   - **Data Center Decision:** The Cyber Risk Monitoring Plan should specifically address the unique threats posed to Data Centers, including physical security breaches, unauthorized access to sensitive data, and disruption of critical infrastructure.

2. Prioritize risks based on potential impact and likelihood.

   - **Business Benefit:** Prioritizing risks allows organizations to focus their efforts on the most critical threats, maximizing the effectiveness of their cybersecurity measures and reducing the potential for costly breaches and disruptions.

   - **Data Center Decision:** Prioritize risks that could lead to data loss, downtime, system failures, or compromise of sensitive information. Consider both the likelihood of an attack and the potential severity of its impact on the Data Center.

3. Allocate resources effectively to address the most critical risks.

   - **Business Benefit:** Effective resource allocation ensures that cybersecurity initiatives have the necessary funding, personnel, and support, increasing the likelihood of successful risk mitigation and ultimately reducing the organization's overall risk exposure.

   - **Data Center Decision:** Allocate resources based on the specific vulnerabilities identified in the Data Center risk assessment. This might include investing in enhanced physical security measures, access controls, robust firewalls, intrusion detection/prevention systems, and data encryption.

4. Implement remediation measures to mitigate identified risks.

   - **Business Benefit:** By proactively addressing identified risks, organizations can prevent or minimize the impact of cyber incidents, preserving business continuity, reputation, and customer trust.

   - **Data Center Decision:** Implement controls and procedures aligned with the Data Center's unique requirements. This includes physical safeguards, strict access protocols, vulnerability patching, incident response plans, and regular security training for staff.

5. Continuously monitor and update the plan to adapt to the ever-changing threat landscape.

   - **Business Benefit:** Regular monitoring and updating of the plan ensure that organizations stay ahead of emerging threats and vulnerabilities, maintaining a strong cybersecurity posture and reducing the likelihood of breaches and disruptions.

   - **Data Center Decision:** Remain vigilant about evolving threats targeting Data Centers. Regularly review threat intelligence, update security technologies, and conduct security audits to identify and address emerging vulnerabilities.

Cyber Risk Monitoring Objectives Defined

The key objectives are to identify and prioritize the threats and vulnerabilities that apply. With these identified items, determine the cost of the cyber risk, the potential remediation for those cyber risks, and the lasting effects of the chosen course of action.

Identify threats: A threat is any activity that represents a possible danger. The 2016 Dyn cyberattack, which affected major websites and services, highlights the significance of direct attacks on infrastructure supporting operations.

Identify vulnerabilities: A vulnerability is a weakness or exposure to threat. The Equifax data breach in 2017, where sensitive personal information of millions of people was compromised, emphasizes the importance of proper monitoring and implementation of access controls and authentication.

Analyze the cyber risks: This involves evaluating the systems related to operations, identifying the threats affecting each component, and identifying the vulnerabilities associated with each component. The 2018 Atlanta ransomware attack serves as an example of the need to analyze cyber risks to protect information and maintain operational stability.

Develop a cyber risk mitigation strategy: Determine appropriate actions to address identified cyber risks, considering their costs and benefits. The NotPetya attack in 2017, which caused significant disruptions to businesses worldwide, demonstrates the need for robust cybersecurity strategies.

Swim Lane Diagram

	Role	Step	Description
01	Assessment Team	•Risk Identification •Risk Analysis	Identify potential cyber risks and vulnerabilities.
02	Risk Managers	•Risk Prioritization	Prioritize risks based on their potential impact and likelihood.
03	Security Team	•Develop and Implement Controls	Develop and implement risk mitigation strategies.
04	Monitoring Team	•Continuous Monitoring	Identify potential cyber risks and vulnerabilities.
05	Incident Response Team	•Incident Response	Respond to incidents promptly and effectively if they occur.
06	Improvement Team	•Review & Update	Conduct post-incident analysis and update CRMP.

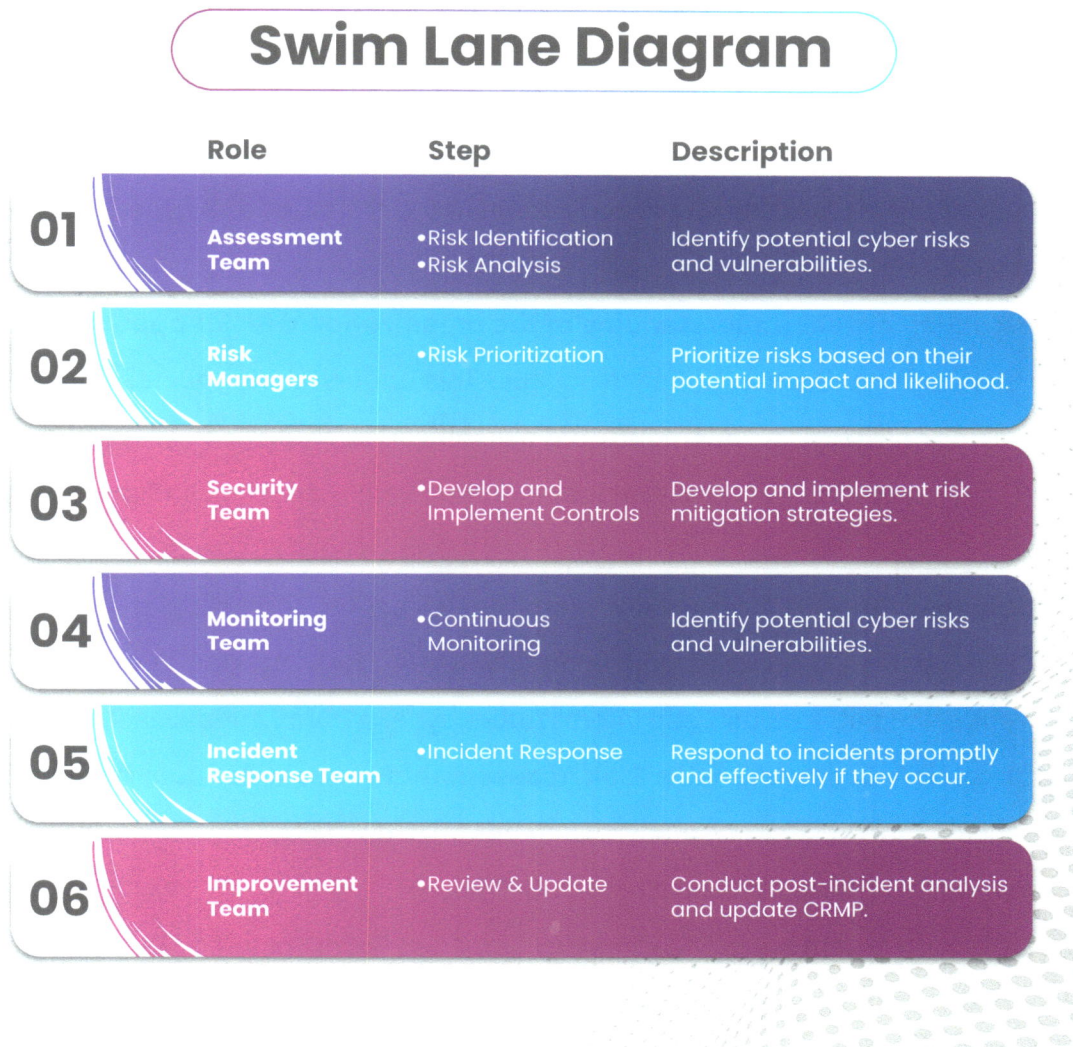

Figure 17: Swimlane Diagram

Key Performance Indicators

Track key performance indicators (KPIs) to measure the effectiveness of cyber risk management efforts. The Cyber Risk Monitoring Plan aims to achieve the following objectives in addition to the main purposes:

1. Establish a comprehensive understanding of the organization's Cyber Risk Exposure, considering both internal and external factors.

   - **Business Benefit:** A holistic view of the organization's risk exposure enables better decision-making and planning, helping to identify and address potential vulnerabilities and areas of improvement in the organization's cybersecurity strategy.

   - **Establish a clear understanding of the organization's Risk Appetite:** How much risk is the organization willing to accept? This understanding guides decision-making on security investments and risk mitigation strategies.

2. Develop a methodology for systematically identifying, assessing, and prioritizing Cyber Risks.

   - **Business Benefit:** A systematic methodology enables organizations to evaluate their cyber risk landscape consistently and effectively, making it easier to identify trends and patterns that can inform future cybersecurity initiatives.

   - **Develop A Comprehensive Risk Assessment Process:** Conduct regular assessments to identify, analyze, and prioritize cybersecurity risks, with a specific emphasis on those unique to the Data Center.

3. Create a clear action plan for addressing Identified Risks, including Remediation Measures, Monitoring, And Reporting.

   - **Business Benefit:** A well-defined action plan promotes accountability and ensures that stakeholders understand their roles and responsibilities in addressing cyber risks, leading to more efficient and effective risk management efforts.

4. Enhance communication and collaboration among various stakeholders, including IT, Security, and Business teams, to ensure a coordinated approach to managing Cyber Risks.

- **Business Benefit:** Improved collaboration and communication among stakeholders can lead to faster identification and resolution of cyber risks, helping to maintain a strong security posture and reducing the potential for costly breaches and disruptions.

- **Define Incident Response Procedures:** Develop a detailed plan for responding to cyber incidents, outlining roles, responsibilities, communication protocols, and Data Center specific recovery procedures.

- **Promote Cybersecurity Awareness:** Conduct regular training for all staff on cybersecurity best practices, including specific training for Data Center personnel focused on physical security, access control, and incident response procedures.

5. Foster a culture of Continuous Improvement in Cybersecurity, with the plan regularly reviewed and updated to reflect changes in the organization's Risk Landscape and technological advancements.

- **Business Benefit:** By embracing a culture of continuous improvement, organizations can stay ahead of emerging threats and vulnerabilities, maintaining a strong cybersecurity posture and reducing the likelihood of breaches and disruptions. This commitment to improvement can also contribute to a positive reputation among customers and partners, demonstrating the organization's dedication to protecting their data.

- **Implement Continuous Monitoring and Reporting:** Establish processes for real-time monitoring of security systems, networks, and Data Center infrastructure to detect anomalies and potential threats.

Cyber Risk Monitoring Plan Purpose and Objectives

The cybersecurity landscape is constantly evolving, with new threats and vulnerabilities emerging regularly. In this context, organizations must be proactive in identifying and managing cyber risks to protect their assets, operations, and reputation. A Cyber Risk Monitoring Plan is a strategic approach to managing cybersecurity threats and mitigating potential consequences. It serves as a roadmap for organizations to effectively prioritize and address cyber risks in a systematic and continuous manner.

This chapter outlines the purpose and key objectives of a Cyber Risk Monitoring Plan, emphasizing its importance in maintaining an organization's overall cybersecurity posture.

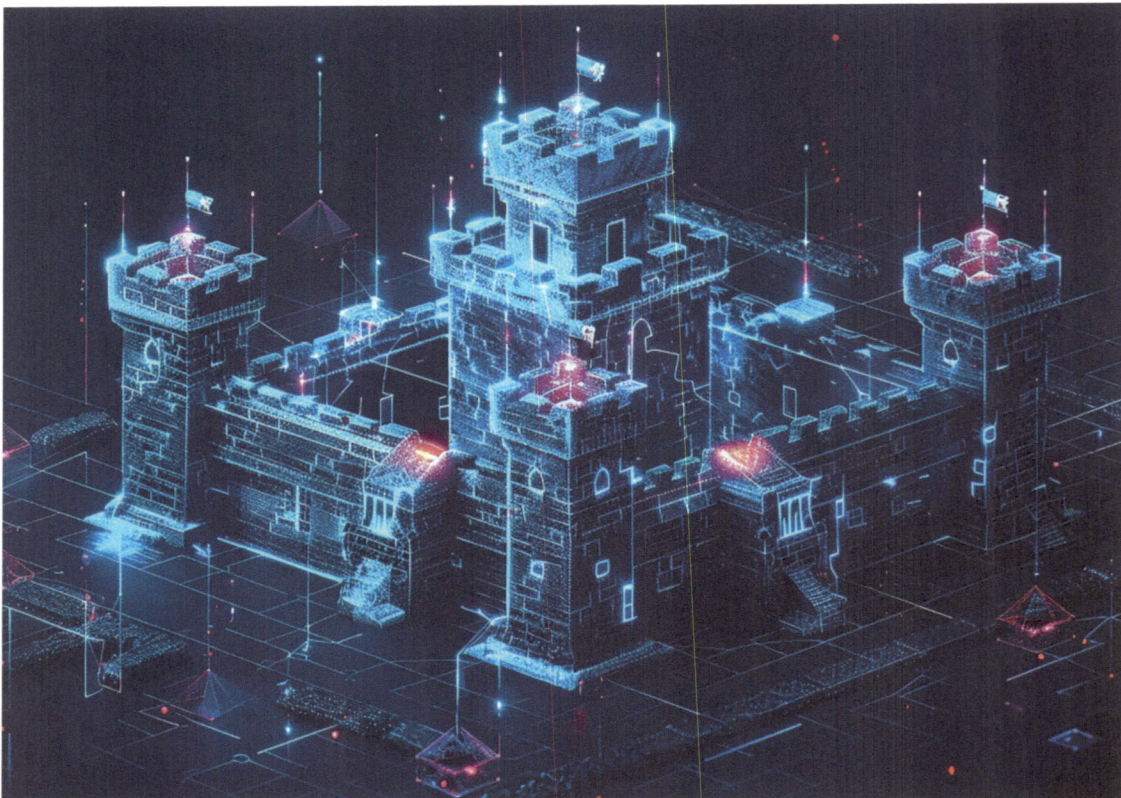

Figure 18: The Digital Watchtower: A Vigilant Eye on Cyber Threats

Purpose and Objectives of a CRMP

A strong CRMP begins with systematic cyber risk identification and analysis. This entails thoroughly assessing the threat landscape, including potential internal vulnerabilities and external threats. Organizations can leverage intelligence operations and advanced analytical tools to gain insights into their risk profile and focus their cybersecurity efforts.

The primary purpose of a Cyber Risk Monitoring Plan is to provide organizations with a structured and proactive approach to:

1. **Understand their unique cybersecurity risk landscape:** This understanding enables informed decisions about cybersecurity investments and strategies, ensuring resources are allocated effectively. For data centers, this means addressing threats like physical security breaches, unauthorized access to sensitive data, and disruption of critical infrastructure.

2. **Prioritize risks based on potential impact and likelihood:** Prioritizing risks allows organizations to focus on the most critical threats, maximizing the effectiveness of their cybersecurity measures. In a data center context, this might involve prioritizing risks that could lead to data loss, downtime, system failures, or compromise of sensitive information.

3. **Allocate resources effectively to address the most critical risks:** Effective resource allocation ensures that cybersecurity initiatives have the necessary funding, personnel, and support. For data centers, this could mean investing in enhanced physical security, access controls, robust firewalls, intrusion detection/ prevention systems, and data encryption.

4. **Implement remediation measures to mitigate identified risks:** By proactively addressing risks, organizations can prevent or minimize the impact of cyber

incidents. For data centers, this could include implementing controls and procedures aligned with their unique requirements, such as physical safeguards, strict access protocols, and regular security training.

5. **Continuously monitor and update the plan:** Regular monitoring and updating ensure organizations stay ahead of emerging threats and vulnerabilities. For data centers, this means remaining vigilant about evolving threats, regularly reviewing threat intelligence, updating security technologies, and conducting security audits.

The CRMP aims to achieve the following key objectives:

- Establish a comprehensive understanding of the organization's cyber risk exposure, both internal and external.
- Develop a methodology for systematically identifying, assessing, and prioritizing cyber risks.
- Create a clear action plan for addressing identified risks, including remediation measures, monitoring, and reporting.
- Enhance communication and collaboration among stakeholders to ensure a coordinated approach to managing cyber risks.
- Foster a culture of continuous improvement in cybersecurity, with the plan regularly reviewed and updated to reflect changes in the risk landscape.

Case Studies and Examples

Real-world incidents like the Equifax data breach and the WannaCry ransomware attack highlight the benefits of a well-executed CRMP in minimizing cyber incidents. By learning from these cases, organizations can better understand the importance of timely vulnerability patching and effective incident response plans.

Roles and Responsibilities

A CRMP's success depends on clearly defined roles and responsibilities across the organization. Collaboration between executive leadership, IT professionals, and other stakeholders is crucial for ensuring a unified and proactive response to cyber threats.

Intelligence Operations Cycle as a Methodology

Integrating the Intelligence Operations Cycle provides a structured methodology for cyber risk management. This proactive, intelligence-driven approach involves planning, collecting data, processing and analyzing information, producing actionable intelligence, and disseminating it to relevant stakeholders.

By following a well-defined methodology and implementing the recommended strategies, organizations can develop a CRMP tailored to their unique needs.

Principles of a Cyber Risk Monitoring Plan

Cyber Risk Focus and Monitoring Plan

The Cyber Risk Monitoring Plan is designed to identify, analyze, and monitor net negative impact cyber risks that can adversely affect an organization's operations. One such example is the 2017 WannaCry ransomware attack, which disrupted operations across various industries, including healthcare, logistics, and manufacturing.

Purpose

The plan's purpose is to establish the objectives of the Cyber Risk Intelligence Monitoring Function. A cyber risk event is any event or condition that can have a positive or negative impact on business operations. The primary mission of the process is to ensure that any cyber risks that can cause undesirable impact on brand, profit, or growth are identified, analyzed (categorized and validated), and presented to the leadership at the right level for action. The vision is to ensure that businesses maintain the ability to perform critical business operations without interruption or excessive cost to the organization.

This plan defines the cyber risks associated with the industry's standard operating model (where one exists) and the organization's current business model. Cyber risks that will be considered and explored will be those that have a net negative impact.

For critical infrastructure, the standard operating model is as follows:

1. **Identify:** The first step in the operating model is to identify the critical infrastructure assets, systems, and key terrain nodes throughout the network that need to be protected. This includes identifying the criticality of the assets, the potential consequences of a disruption, and the threats that the assets are exposed to.

- **Cyber Risk:**
  1. **Data Classification:** Identifying the most sensitive/critical data is essential for prioritizing protection efforts.
  2. **Mapping Dependencies:** Understanding complex interdependencies between infrastructure elements is crucial, as disruptions upstream can have cascading effects.
  3. Identify assets, systems, and key terrain nodes, aligning them with your business objectives and threat assessments.
- **Data Center:**
  - ○ **Environmental Risks:** Beyond cyberattacks, data centers face threats like power outages, fires, and natural disasters. These should be factored into asset identification.
  - ○ **Vendor Relationships:** Data centers may house equipment or data belonging to other entities. These interdependencies must be mapped for accurate risk assessment.

2. **Protect:** The second step is to protect the critical infrastructure assets by implementing appropriate security measures. This includes physical security measures, such as access controls and surveillance, as well as cybersecurity measures, such as firewalls, intrusion detection systems, and encryption.
   - **Cyber Risk**
     - ○ **Zero Trust Architecture:** Mention this as a framework increasingly vital for critical infrastructure, where assuming any internal component is safe is dangerous.

      ○ **Third-Party Risk Management:** Emphasize that protection includes rigorous auditing of vendors who have access to or provide services for your critical systems.

- **Data Center**

      ○ **Physical Redundancies:** It is not enough to protect the data itself; power supplies, cooling systems, etc. also need backup and failsafes.

      ○ **Employee Training:** Authorized personnel can still make mistakes. Training on safe maintenance and change management procedures is crucial.

3. **Detect:** The third step is to detect potential threats and incidents by monitoring the critical infrastructure assets and networks. This includes using telemetry data from various sources, such as network traffic, system logs, and security sensors, to detect anomalies and potential security incidents.

- **Cyber Risk**

      ○ **Behavior-Based Analytics:** Critical infrastructure is a prime target for advanced attackers. Tools that detect unusual activity patterns, not just known flaws, are essential.

      ○ **Intrusion Testing:** Proactive "red teaming" helps find vulnerabilities before adversaries do.

- **Data Center**

      ○ **Centralized Monitoring:** Data from sensors, logs, and even employee observation reports on physical spaces needs to be aggregated for a truly holistic view of potential problems.

- **Early Warning Systems:** Canaries in the coal mine – automated alerts for things like subtle temperature changes can prevent a minor issue from cascading.

4. **Respond:** The fourth step is to respond to security incidents in a timely and effective manner. This includes developing incident response plans, establishing communication protocols, and conducting regular training and exercises to ensure that response procedures are effective.

- **Cyber Risk**

  - **Isolate and Segment:** Critical infrastructure attacks demand rapid containment capabilities to protect the most essential functions even if parts of the network are compromised.

  - **Collaboration with Law Enforcement:** Mention the necessity of having established relationships in place before a major attack, as the response may involve specialized expertise.

- **Data Center**

  - **Chain of Custody Protocols:** If an incident results in legal action, the meticulous handling of physical evidence from the facility becomes critically important.

  - **Communication Plan:** This includes notifying affected clients or customers in a timely way that follows regulatory mandates for your industry.

5. **Recover:** The final step is to recover from security incidents and restore critical infrastructure operations. This includes developing recovery plans, establishing backups and redundancies, and conducting regular testing to ensure that recovery procedures are effective.

- **Cyber Risk**
    - **Immutable Backups:** Ransomware attacks cripple recovery if backups are also encrypted. Technologies for ensuring certain backup sets cannot be altered are important to discuss.
    - **Lessons Learned Processes:** Thorough attack analysis is vital so that recovery is not just a return to how things were, but an improvement of defenses to prevent recurrence.

- **Data Center**
    - **Alternate Sites:** Depending on risk tolerance, having "hot" or "cold" backup data centers may be considered essential to recovery.
    - **Disaster Recovery Drills:** It is not enough to have a plan on paper; regular simulations ensure staff can restore critical functionality within acceptable time frames.

Overall, the operating model for Critical Infrastructure is designed to provide a comprehensive approach to protecting critical infrastructure assets and networks. It is a continuous process that involves ongoing monitoring, assessment, and improvement to ensure that critical infrastructure remains secure and resilient in the face of evolving threats and risks.

Considerations

The resulting documentation and artifacts/documents may include a listing of activities and responsibilities required to perform, record, and monitor the prioritization, mitigation, transference, avoidance, or acceptance of these cyber risks. The purpose of these artifacts is to guide the orchestration of the plan and the management of the resulting implementation for our primary customers (the organization).

This document should be updated throughout the lifecycle of operations. The intent is for the management and leadership teams to review and respond to the real-time cyber risks with this document and the policy and procedures outlined in it as owners and sponsors who will ensure implementation and follow-through.

Cyber Risk Monitoring Methodology

Operations

Developing an effective Cyber Risk Monitoring Plan requires a robust methodology that provides a structured approach to identifying and mitigating cyber risks. In this chapter, we will focus on the Intelligence Operations Cycle as a methodology for managing cyber risks. The Intelligence Operations Cycle is a continuous process that involves collecting, analyzing, and disseminating information to support decision-making and reduce risks.

The Process

This chapter provides an overview of the Cyber Risk Monitoring Methodology, which is essential for organizations to effectively identify, assess, and manage cyber risks. The methodology consists of several key steps that help organizations proactively address potential threats and vulnerabilities, ensuring a secure and resilient infrastructure. Each step includes an explanation of its purpose and business benefits, emphasizing the importance of a comprehensive approach to cyber risk management.

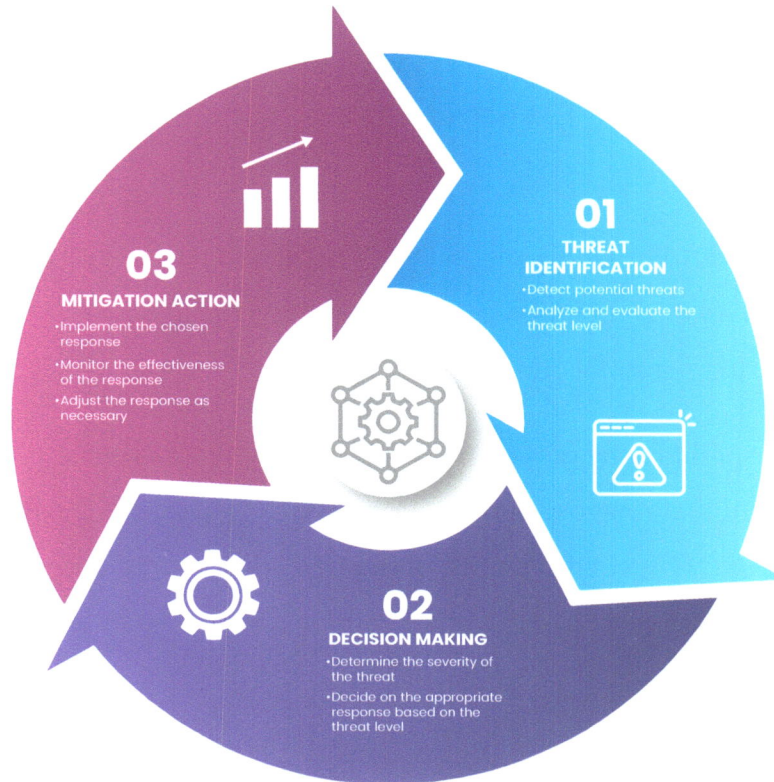

Figure 19: Augmentation

1. Cyber Risk Monitoring Methodology

   - Identify Threats

   - Identify Vulnerabilities

   - Analyze Cyber Risks

   - Evaluate and Define Remediations

   - Conduct Cost-Benefit Analysis

- Document, Track, and Monitor Implementation
- Create a Plan of Action and Milestones (POAM)

Identify Threats

Purpose: To recognize potential cyber threats that may target an organization's systems, assets, or data. Threat intel must be translated into specific monitoring and mitigation actions.

Business Benefit: Early identification of threats enables organizations to prioritize resources and focus on the most relevant risks, reducing the likelihood of successful attacks and minimizing potential damage.

Definition: A cyber threat is any potential malicious activity that seeks to unlawfully access data, disrupt digital operations, or damage information. ("What Is a Cybersecurity Threat: Exploring Cyber Threat") These threats can come from various sources, including individual hackers, criminal organizations, or even state actors, aiming to steal information, extort money, or disrupt digital life.

Approach: The process of identifying threats involves several key steps:

Threat Intelligence Gathering: This includes monitoring threat intelligence feeds, analyzing trends in cybercrime, cyberterrorism and cyber state and non-state actors and staying informed about the latest hacking techniques. This proactive approach enables organizations to anticipate potential threats before they manifest.

System and Network Monitoring: By continuously monitoring the system and activities, organizations can detect unusual patterns that may indicate a threat. This includes monitoring for signs of phishing, malware, ransomware, and other forms of attacks.

Vulnerability Scanning and Penetration Testing: Regularly scanning for vulnerabilities in software and systems allows organizations to identify weaknesses that could be exploited by threats. Penetration testing takes this a step further by attempting to exploit these vulnerabilities in a controlled environment, simulating an attack to assess the system's defenses.

Leveraging Threat Intelligence:

Threat intelligence is the lifeblood of proactive cyber defense. It is not just data; its actionable insights distilled from a sea of information. Think of it as your organization's early warning system, providing a heads-up on the tactics, techniques, and procedures (TTPs) of threat actors targeting your industry or even you specifically.

Why is this crucial? Because knowledge is power. By understanding the threats lurking on the horizon, you can tailor your defenses to counter them before they strike. It is the difference between reacting to a breach after the damage is done and proactively hardening your systems to prevent it from happening in the first place.

Where does this intel come from? There is a wealth of sources at your disposal. Industry-specific ISACs (Information Sharing and Analysis Centers) provide a collaborative space for sharing threat data and best practices. Government agencies like CISA issue alerts on emerging threats, and security vendors publish detailed reports on vulnerabilities and attack trends. Even open-source feeds, with careful vetting, can offer valuable insights. To harness the power of threat intelligence, you need the right tools. Think of threat feeds as your radar, constantly scanning the digital landscape for signs of trouble. Aggregation platforms help you make sense of the vast amounts of data, turning raw information into actionable intelligence.

Threat intelligence is not just for the big players. Every organization, regardless of size or industry, can benefit from this proactive approach to cybersecurity. By staying informed and adapting your defenses, you can build a more resilient future for your business in the face of ever-evolving threats.

Example: The 2016 Dyn cyberattack, a sophisticated Distributed Denial of Service (DDoS) attack, leveraged tens of millions of IP addresses to take down the Dyn domain name system, affecting major websites and services. This attack highlighted the importance of robust threat intelligence and network monitoring to detect and mitigate such widespread threats.

Best Practice in Operations:
- Conduct regular vulnerability scans and penetration tests to identify potential weaknesses in the system.
- Utilize threat intelligence feeds and monitoring tools to identify emerging threats and attack trends.
- Implement intrusion detection and prevention systems to identify and block malicious traffic.
- Training is also a best practice, ensuring that cyber security personnel and threat analyst are trained properly.

Identify Vulnerabilities

Purpose: To uncover weaknesses in an organization's systems, processes, and policies that could be exploited by threat actors.

Business Benefit: Identifying vulnerabilities allows organizations to address weaknesses and implement security measures before they can be exploited, enhancing overall security posture, and reducing the risk of breaches.

Diligent vulnerability identification allows organizations to:

- Patch or mitigate identified weaknesses before they can be exploited.

- Enhance their defenses against potential attacks.

- Maintain trust and reliability among customers and stakeholders by safeguarding sensitive information.

- Enhance and enrich the systems and tools in place.

Definition: A **vulnerability** refers to a flaw or weakness in a system's design, implementation, operation, or management that could be exploited to violate the system's security policy. Vulnerabilities can exist in software, hardware, or organizational processes.

Approach: To effectively identify vulnerabilities, organizations should:

1. **Conduct Regular Security Assessments:** These assessments help to uncover weaknesses in the organization's security posture, including misconfigurations, outdated software, and weak security policies. These policies include facility handover without coordinated access management for network and compute assets. Failure to control modules and equipment.

2. **Utilize Automated Vulnerability Scanning**

3. **Tools:** These tools can scan systems and networks for known vulnerabilities, providing a quick and efficient means of identification.

4. **Human: Assessments** can be manual or automated. Assessments and scans can be human operated, or tool based.

5. **Engage in Continuous Monitoring:** Implementing continuous monitoring strategies can help detect new vulnerabilities as they arise, especially those that may be introduced through software updates or new system integrations.

Example: The Equifax data breach in 2017, where sensitive personal information of millions was compromised, was the result of exploiting a known vulnerability in the Apache Struts web framework. This incident underscores the critical need for timely vulnerability identification and patch management.

Best Practice in Operations:

- Implement software patching policies to address known vulnerabilities in applications and operating systems.
- Conduct regular security assessments to identify security gaps in the infrastructure.
- Utilize systems segmentation to limit the impact of a breach and prevent lateral movement.

Analyze Cyber Risks

Purpose: The purpose of analyzing cyber risks is to evaluate the potential impact and likelihood of threats and vulnerabilities within an organization's digital landscape, taking into account its unique context and risk tolerance. This analysis is pivotal in prioritizing risks, guiding decision-making processes, and optimizing cybersecurity strategies to protect organizational assets.

Definitions:

- **Cyber Risk Analysis** involves the systematic examination of threat scenarios to evaluate the potential impacts on an organization and determine the likelihood of occurrence. It encompasses assessing the vulnerabilities that could be exploited by these threats and the existing controls' effectiveness in mitigating such risks.

- **Risk Tolerance:** The level of risk an organization is prepared to accept before action is deemed necessary. ("The maximum level of risk that an organization is - Course Hero") It reflects the organization's capacity to withstand risks in pursuit of its objectives.

- **Context:** The specific circumstances or environment in which cyber risks are assessed. This includes the organization's assets, threat landscape, business operations, and strategic goals.

A comprehensive risk analysis facilitates informed decision-making regarding resource allocation and risk mitigation strategies. By understanding *the severity and likelihood* of various cyber risks, organizations can focus their efforts and investments on areas with the highest potential impact, thereby enhancing their cybersecurity posture and safeguarding critical assets.

The Approach: The analysis process begins with the identification of potential threats and vulnerabilities, followed by an assessment of the impact and likelihood of each identified risk. This includes:

- **Evaluating the potential impact** of each risk on the organization's operations, reputation, and data integrity.
- **Determining the likelihood** of each risk materializing based on current threat intelligence, internal security posture, and historical data.
- **Prioritizing risks** based on their calculated severity and likelihood, allowing for targeted mitigation efforts.

Example: A pertinent example of cyber risk analysis in action is the response to the **NotPetya malware attack in 2017**, which caused widespread disruption and financial losses. Organizations that had conducted thorough risk analyses and prioritized their cybersecurity efforts were able to respond more effectively to the incident, highlighting the value of preparedness and informed risk management.

Best Practice in Operations:

- **Conduct regular risk assessments:** Frequent evaluations ensure that emerging threats and new vulnerabilities are identified and addressed promptly.

- **Utilize SIEM tools:** Security Information and Event Management (SIEM) systems play a critical role in analyzing system logs and detecting suspicious activities, offering real-time insights into potential threats.

- **Implement EDR solutions:** Endpoint Detection and Response (EDR) tools are essential for identifying and responding to threats at the device level, providing an additional layer of security.

Analysis Types and Their Applications

Type of Analysis	Focus	Primary Metric(s)	Complexity	Source Organization	Equations	Common Applications
Cost-Benefit Analysis (CBA)	Economic efficiency	Net Present Value (NPV), Benefit-Cost Ratio (BCR)	Moderate -High	Public agencies, private sectors, consultancies	$NPV = \Sigma$ (Benefits - Costs) / $(1+r)^t$ $BCR = \Sigma$ Benefits / Σ Costs	Projects with quantifiable financial costs &benefits
Risk-Benefit Analysis (RBA)	Balancing potential harms & benefits	Qualitative & semi-quantitative assessments of likelihood & severity	Variable	Regulatory agencies, industry, public health organizations	No universal equation: risk matrices oftenused	Decisions involving potential safety, health, or environmental risks
Fiscal Impact Analysis (FIA)	Budgetary impact on local governments	Net fiscal impact (revenue vs. costs)	Moderate	Local governments, economic development agencies	Net Fiscal Impact = Projected Revenues - Projected Costs	Land use, infrastructure development, economic development projects
Social Impact Analysis (SIA)	Community & societal impact	Qualitative and quantitative indicators	High	Community organizations, NGOs, government agencies, consultants	No universal equation: impact frameworks vary	Projects with social, cultural, environmental, & economic implications

Figure 20: Analytical Frameworks for Decision-Making: Beyond Cost-Benefit Analysis

Evaluate and Define Remediations

Purpose: The primary goal in this step is to design and prioritize remedial actions derived from the risk analysis outcomes, ensuring a balanced evaluation of the cost-effectiveness of each proposed solution. This step is pivotal in shaping the cybersecurity framework to address identified vulnerabilities and threats effectively.

Business Benefit: Evaluating and defining remediation strategies offers substantial business advantages by ensuring that organizations deploy the most beneficial cybersecurity measures tailored to their unique needs. This approach not only maximizes the return on investment in cybersecurity but also significantly diminishes the residual risks, thereby enhancing the security posture and resilience against cyber threats.

Definitions:

- **Remediation** involves actions taken to mitigate or eliminate the vulnerabilities within an organization's IT environment, thus reducing the potential for exploitation by cyber threats. Effective remediation is a targeted response to the specific risks identified during the cyber risk analysis process, aiming to fortify the organization's defenses against future attacks.

- **Residual risk** refers to the amount of risk that remains after security measures have been applied. ("Difference between Residual Risk and Inherent Risk") It is the level of risk that an organization accepts because it's either too costly or impractical to reduce it further through additional controls. In the context of cybersecurity, it involves understanding that no security measure is 100% effective, and some level of risk will always be present. Effective risk management strategies aim to reduce risk to an acceptable level, not to eliminate it entirely. The concept is crucial in cybersecurity planning and decision-making, guiding organizations on

107

where to allocate resources most effectively to protect against cyber threats while acknowledging that not all risks can be fully mitigated.

Best Practice in Operations:

- Implement access controls to limit privileges and permissions to only those necessary for job functions.
- Implement multi-factor authentication to prevent unauthorized access to systems and data.
- Utilize encryption to protect sensitive data in transit and at rest.

Approach: The approach to evaluating and defining remediations encompasses a detailed analysis of each identified risk, considering both the potential impact on the organization and the feasibility of the proposed mitigation strategies. This process involves:

- **Assessment of Remediation Options:** Exploring various security measures that could mitigate identified risks, including technological solutions, policy changes, and training programs.
- **Cost-Benefit Analysis:** Weighing the costs associated with implementing each remediation against the potential benefits in terms of reduced risk exposure.
- **Prioritization:** Ranking the remediation actions based on their urgency, impact, and effectiveness in mitigating risks.

Example: A notable example is the SolarWinds attack, where vulnerabilities in the software supply chain were exploited. Organizations affected by this incident had to evaluate and define a set of remediations that included applying patches, enhancing monitoring of network traffic, and revising access controls. This case underscores the importance of timely and targeted remediations to address vulnerabilities and prevent the exploitation of organizational assets.

Best Practices in Operations (Referenced from Cyber Defense Risk Monitoring Methodology and Intelligence Operations - 2024):

- **Implement Access Controls:** Restrict access rights to systems and data to the minimum necessary for job functions, minimizing the potential attack surface.

- **Multi-factor Authentication (MFA):** Deploy MFA to add an additional layer of security for system and data access, significantly reducing the risk of unauthorized access due to compromised credentials.

- **Encryption:** Apply encryption techniques to protect sensitive data in transit and at rest, ensuring data integrity and confidentiality.

- **Regular Security Assessments:** Conduct periodic security assessments to identify new vulnerabilities and ensure that remediation measures are effectively addressing identified risks.

Conduct Cost-Benefit Analysis

Purpose: The purpose of conducting a cost-benefit analysis in cybersecurity is to weigh the financial investment required for implementing security measures against the anticipated benefits in terms of risk reduction. This evaluation is essential for ensuring that resources are allocated efficiently, focusing on mitigations that offer the most significant return on investment and enhance the organization's security posture.

Business Benefit: Performing a cost-benefit analysis provides tangible value by enabling organizations to prioritize cybersecurity initiatives that deliver the most substantial impact in mitigating risks. This strategic approach ensures that investments are made in areas that significantly reduce vulnerabilities and potential damage from cyber threats, thereby optimizing the use of limited resources and maximizing cybersecurity effectiveness.

Definition: Cost-benefit analysis in cybersecurity involves quantifying the costs associated with implementing specific security measures (such as software tools, training, or infrastructure changes) and comparing these costs to the benefits derived from reduced risk exposure and potential impact of cyber incidents. This analysis helps in identifying and prioritizing the security measures that offer the best value.

Approach: Organizations should adopt a structured approach to conduct cost-benefit analyses by:

- Identifying potential security investments and the risks they aim to mitigate.
- Estimating the costs associated with each investment, including initial deployment, ongoing maintenance, and any potential operational impacts.
- Assessing the benefits of each investment in terms of risk reduction, compliance with regulatory requirements, and improvement in security posture.

- Prioritizing investments based on their net benefits and alignment with the organization's risk tolerance and strategic objectives.

Example: Utilizing risk management frameworks such as NIST or ISO provides a structured methodology for conducting cost-benefit analyses. These frameworks offer guidelines for assessing risks, evaluating the effectiveness of controls, and determining the cost-effectiveness of risk mitigation strategies.

Best Practice in Operations:

- **Utilize risk management frameworks** (such as NIST or ISO or PMI) to guide the cost-benefit analysis process, ensuring a comprehensive evaluation of potential security investments (Cyber Defense Risk Monitoring Methodology).

- **Conduct regular security awareness training** for employees to minimize the risk of human error leading to security incidents. Training is a cost-effective measure that significantly reduces the likelihood of successful phishing attacks and other social engineering tactics.

- **Employ threat modeling techniques** to assess the potential impact of security incidents and prioritize investments in security measures. Threat modeling helps in understanding the most critical vulnerabilities and the effectiveness of proposed mitigations in protecting against specific attack scenarios.

Document, Track, and Monitor Implementation

Purpose: The primary objective of documenting, tracking, and monitoring implementation is to maintain a systematic record and oversight of the progress in addressing cyber risks, ensuring the effective application and persistence of security measures within an organization. This process is vital for verifying that risk mitigation strategies are not only implemented but also maintained over time to protect against evolving threats.

Business Benefit: Engaging in rigorous documentation and tracking delivers several key benefits, including bolstering an organization's cybersecurity posture, supporting compliance with relevant regulations, and enhancing stakeholder confidence. It demonstrates a proactive and responsible approach to managing cyber risks, which can be crucial for maintaining trust with customers, partners, and regulatory bodies.

Definitions:

- **Documentation** involves creating detailed records of all cyber risk mitigation actions, including plans, implementation steps, and outcomes.
- **Tracking** refers to the continuous monitoring of these actions to ensure progress and adherence to schedules.
- **Monitoring** encompasses the ongoing surveillance of implemented measures to verify their effectiveness and detect any deviations or failures promptly.

Approach (Based on Cyber Defense Risk Monitoring Methodology):

- Implement a **security incident response plan** that outlines procedures for documenting and addressing security incidents, ensuring a structured response to threats (Cyber Defense Risk Monitoring Methodology).

- Use **Security Information and Event Management (SIEM) tools** to continuously monitor the network for suspicious activities and anomalies, providing real-time insights into potential security breaches.

- Conduct **regular security audits** to assess compliance with industry standards and internal policies, identifying areas for improvement and ensuring that documentation is kept up to date and accurately reflects the organization's security posture.

Example: A practical example of this approach in action can be seen in organizations that have established a **comprehensive cyber risk register**. This register documents all identified risks, their mitigation strategies, implementation statuses, and monitoring outcomes. Regular audits are conducted to assess the effectiveness of these strategies and adjustments are made based on the audit findings, with all changes meticulously documented and tracked.

Best Practice in Operations:

- **Regular Review and Update:** Ensure that all cyber risk documentation is regularly reviewed and updated to reflect the current threat landscape and the organization's evolving security posture.

- **Stakeholder Engagement:** Engage relevant stakeholders in the documentation process to ensure comprehensive coverage of all aspects of cyber risk management.

- **Integration with Incident Response:** Seamlessly integrate documentation and tracking practices with the organization's incident response plan to ensure swift and effective action in the face of new threats.

Create a Plan of Action and Milestones (POAM)

Purpose: The development of a Plan of Action and Milestones (POAM) is aimed at laying down a detailed, actionable strategy to mitigate identified cyber risks. This plan outlines specific steps to be taken, assigns responsibilities, and sets timelines for each action, ensuring a coordinated response to cyber threats.

Definition: A POAM is defined as a comprehensive, structured document that identifies high-risk vulnerabilities within an organization and outlines actions to address these issues. It includes timelines for implementation, resource allocation, and responsibilities for each task, providing a roadmap for risk mitigation efforts.

Business Benefit: A well-crafted POAM offers numerous benefits, including keeping cybersecurity efforts on track and ensuring timely mitigation of risks. It fosters a proactive risk management culture, enhancing an organization's ability to respond to and recover from cyber incidents, thereby preserving operational continuity and safeguarding against potential financial and reputational damage.

Approach: The approach to creating a POAM involves several key steps:

- **Risk Identification and Prioritization:** Utilizing findings from cyber risk assessments to identify and prioritize risks based on their potential impact and likelihood.
- **Action Item Development:** For each identified risk, developing specific, measurable actions that can mitigate the risk to an acceptable level.
- **Assignment of Responsibilities:** Designating team members or departments responsible for implementing each action item.

- **Timeline Establishment:** Setting realistic deadlines for the completion of each mitigation action to ensure timely risk reduction.

Example: In the case of the SolarWinds attack, organizations affected could have developed a POAM that outlined steps to isolate compromised systems, apply patches, and conduct thorough security audits. The POAM would detail the responsible parties for each action, resources required, and a timeline for execution to quickly mitigate the impact and prevent future breaches.

Best Practices in Operations (Referenced from Cyber Defense Risk Monitoring Methodology):

- **Utilization of a Ticketing System:** To effectively track progress on security remediations, ensuring that each action item is followed through to completion.
- **Regular Security Risk Assessments:** To identify new risks and potential mitigations, ensuring the POAM remains relevant and up to date with the evolving threat landscape.
- **Vulnerability Management Tools:** To track and prioritize patching efforts, ensuring vulnerabilities are addressed in a timely manner, thereby reducing the window of opportunity for attackers.

By following these guidelines, organizations can ensure that their POAM is not only comprehensive but also actionable and aligned with their overall cybersecurity strategy, ultimately enhancing their resilience against cyber threats.

Cyber Risk Documentation, Case Studies, and the Path Forward

Documenting and Reporting for Visibility and Action

Documentation and reporting are the cornerstones of an effective Cyber Risk Monitoring Plan (CRMP). Creating a comprehensive Cyber Risk List to record identified risks, assessments, and mitigation measures is crucial. Accurate documentation keeps stakeholders informed about the evolving risk landscape and how your organization is responding. Regular updates and reports promote transparency, ensuring that everyone understands their roles in risk management.

Case Studies: Lessons from the Front Lines

Real-world incidents like the Equifax data breach and WannaCry ransomware attack provide stark reminders of why robust cyber risk management is so vital. These case studies clearly show the severe impact of inadequate defenses and underscore the immense value of proactive risk identification and analysis. By analyzing these incidents, organizations can learn from the mistakes of others and strengthen their own strategies.

Conclusion and Future Directions

Identifying and analyzing cyber risks lays the foundation for a more secure organizational future. ***Following a structured methodology, leveraging intelligence operations, and prioritizing accurate documentation*** allows organizations to better manage an ever-changing cyber threat landscape. However, the cybersecurity journey is ongoing. As new threats emerge, so must organizations adapt their responses. This requires continuous vigilance, adjusting strategies, and a commitment to continual improvement.

Key Takeaways

- **Documentation is Power:** A well-maintained Cyber Risk List fosters transparency and drives action.

- **Learn from the Past:** Case studies offer invaluable lessons to guide your own risk management efforts.

- **The Future Is Dynamic:** Embrace continuous improvement to stay ahead of evolving cyber threats and build a resilient digital future.

Tabletop Exercises: Simulating Real-World Threats

Tabletop exercises are a powerful tool for assessing an organization's preparedness for cyberattacks. By simulating realistic scenarios in a controlled environment, these exercises allow teams to practice their response procedures, identify weaknesses, and improve communication and decision-making.

A well-designed tabletop exercise can:

- Uncover hidden vulnerabilities and gaps in security controls.
- Evaluate the effectiveness of incident response plans and procedures.
- Test communication and coordination between different teams.
- Identify training needs for employees and stakeholders.
- Build confidence and preparedness in the face of cyber threats.

To get started with tabletop exercises, download our free sample scenario, which provides a step-by-step walkthrough of the process, https://rb.gy/czt93z

Cyber Risk Assessment Outcomes

Cyber risk assessments are a critical component of any organization's cybersecurity strategy. The goal of these assessments is to identify potential risks and vulnerabilities and determine the impact they could have on the organization's operations. Once the risks have been identified, it is important to categorize them based on their severity and determine the appropriate actions to take to mitigate them.

In this chapter, we will discuss the outcomes of cyber risk assessments and the various categories of cyber risks.

Risk Categories

There are four categories of cyber risks that organizations should be aware of: Improve, Monitor, Tolerate, and Operate.

1. **Improve:** Risks that fall into this category require immediate action. These are high-impact risks that could cause significant damage to the organization if left unchecked. Examples of risks in this category include unpatched vulnerabilities, weak passwords, and misconfigured systems. The appropriate action for these risks is to implement controls to eliminate or reduce them.

2. **Monitor:** Risks in this category are less severe than those in the Improve category, but still require ongoing monitoring. Examples include unusual network activity, failed login attempts, and suspicious email activity. The appropriate action for risks in this category is to implement monitoring controls to detect any potential threats.

3. **Tolerate:** Risks in this category are low-impact and can be accepted without any significant impact on the organization's operations. Examples include outdated software and low-risk user behavior. The appropriate action for risks in this category is to accept them and not allocate significant resources to mitigate them.

4. **Operate:** Risks in this category are those that are necessary for the organization to operate. Examples include remote access and user privileges. The appropriate action for risks in this category is to implement controls to ensure that they are managed properly.

Risk Assessment Outcomes

After conducting a cyber risk assessment, there are three possible outcomes: Accept, Mitigate, or Transfer.

1. **Accept:** Risks that fall into the Tolerate and Operate categories can be accepted without further action. These risks are considered low-impact and can be managed with existing controls.

2. **Mitigate:** Risks that fall into the Improve and Monitor categories require mitigation. Mitigation strategies can include implementing new controls, updating existing controls, or applying patches and updates.

3. **Transfer:** Risks can be transferred through insurance or third-party contracts. This is often done for risks that are too expensive or difficult to mitigate.

Risk Management Documentation

One critical component of cyber risk management is documentation. Documentation provides a record of the organization's risk management efforts and can be used to demonstrate compliance with regulatory requirements.

The Cyber Risk List is a key document in risk management. It lists all identified risks, their associated categories, and their risk assessment outcomes. Each risk should be assigned an owner who is responsible for implementing the appropriate mitigation strategy.

In addition to the Cyber Risk List, other documentation may include risk assessment reports, incident response plans, and security policies and procedures.

Conclusion

Cyber risk assessments are an essential part of any organization's cybersecurity strategy. By categorizing risks and determining the appropriate actions to take, organizations can better protect themselves from cyber threats. The outcomes of risk assessments, including the Cyber Risk List and risk management documentation, provide a record of the organization's risk management efforts and can be used to demonstrate compliance with regulatory requirements.

Meet Ryan Williams Sr., Cyber Influencer and Thought Leader

Figure 21: Photo of Ryan Williams Sr.

Contribution by Ryan Williams Sr., Cybersecurity Professional, Podcaster, Retired Air Force Veteran

"As cyber threats continue to evolve, so too must our approaches to monitoring and managing these risks. The NIST CSF is well-positioned to adapt to these changes, with ongoing updates that incorporate emerging threats and technologies. Continuous improvement and adaptation are key to maintaining a strong cybersecurity posture, and the NIST CSF provides a solid foundation for this endeavor.

Cybersecurity Frameworks and Standards

Overview of NIST Framework

Introduction to NIST CSF:

The NIST Cybersecurity Framework (CSF) serves as a guideline and tool to help organizations of all sizes and sectors better manage cybersecurity risks. Its core purposes are:

- **Standardization:** The CSF provides a common language and structure for describing cybersecurity practices, allowing organizations to assess their posture, communicate their needs, and benchmark against others.

- **Risk Focus:** This risk-based approach prepares organizations to combat threats like ransomware **Management**, data theft, and insider attacks – all of which we have discussed. The framework emphasizes understanding, assessing, and proactively mitigating cyber risks. It aims to shift organizations away from purely reactive approaches.

- **Adaptability:** The CSF is designed to be flexible and adaptable to different organizations, considering their unique risk profiles, regulatory requirements, and resources.

- **Prioritization:** The framework guides organizations in prioritizing cybersecurity investments and actions based on their most critical risks.

The NIST Cybersecurity Framework (CSF) establishes a foundational approach to managing cybersecurity risks through its core functions: Identify, Protect, Detect, Respond, and Recover.

- **Identify:** This function focuses on understanding an organization's cybersecurity landscape. A CRMP supports this through ensuring an understanding of risk and assessing current posture, which helps identify critical assets, vulnerabilities, and relevant threats.

- **Protect:** This involves implementing safeguards to limit the impact of potential cyber incidents. An effective CRMP emphasizes remediation strategies, technology tools, and access controls, contributing to fulfilling the Protect function.

- **Detect** This function emphasizes continuous monitoring of systems and networks for signs of intrusion or compromise. The CRMP's focus on threat intelligence, vulnerability scanning, and log analysis strengthens an organization's ability to detect potential threats early.

- **Respond:** When incidents occur, a timely and effective response is crucial. The CRMP's inclusion of incident response planning aligns with this function, helping organizations contain breaches and minimize damage.

- **Recover:** This focuses on restoring operations and capabilities following a cybersecurity event. The CRMP's development of disaster recovery and business continuity planning ensures organizations can recover effectively and maintain resilience.

A robust and effective Cyber Risk Monitoring Plan directly supports all these functions.

CSF and The Cyber Risk Methodology

Complementary, Not Competitive

The NIST Cybersecurity Framework (CSF) serves as the overarching blueprint for an organization's cybersecurity program, outlining the essential goals and functions. The Cyber Risk Monitoring Plan (CRMP), on the other hand, provides a detailed wiring diagram – the specific steps, tools, and tactics for identifying, assessing, and continuously monitoring cyber risks. Together, they provide both the strategic vision and the practical execution needed for robust cybersecurity.

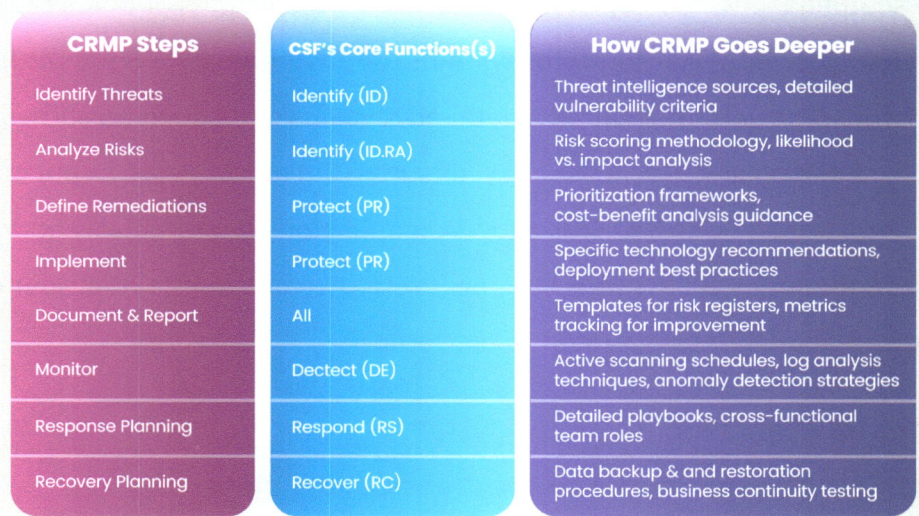

CRMP Steps	CSF's Core Functions(s)	How CRMP Goes Deeper
Identify Threats	Identify (ID)	Threat intelligence sources, detailed vulnerability criteria
Analyze Risks	Identify (ID.RA)	Risk scoring methodology, likelihood vs. impact analysis
Define Remediations	Protect (PR)	Prioritization frameworks, cost-benefit analysis guidance
Implement	Protect (PR)	Specific technology recommendations, deployment best practices
Document & Report	All	Templates for risk registers, metrics tracking for improvement
Monitor	Dectect (DE)	Active scanning schedules, log analysis techniques, anomaly detection strategies
Response Planning	Respond (RS)	Detailed playbooks, cross-functional team roles
Recovery Planning	Recover (RC)	Data backup & and restoration procedures, business continuity testing

Figure 22: Aligning Cyber Risk Monitoring with the NIST Cybersecurity Framework (CSF)

CRMP as a CSF Subset

Examples of Synergies

125

Incident Response Synergy: CRM's emphasis on continuous monitoring significantly enhances the CSF's Respond function. Detailed network logs, anomaly detection, and timely alerts generated through a CRMP provide security teams with the early warning signs of a potential breach. This allows for faster containment, minimizing damage and reducing recovery time. The CRMP's well-defined incident response playbooks further streamline the response process, ensuring timely and coordinated actions that effectively neutralize the threat.

Vulnerability Management Synergy: CRMP's integrated approach to vulnerability management supports both the Protect and Detect functions of the CSF. Frequent vulnerability scanning, informed by the CRMP's threat intelligence, enables proactive patch management and the hardening of systems, strengthening the organization's defensive posture (Protect). The CRMP's detailed analysis of discovered vulnerabilities helps prioritize those posing the highest risk, ensuring the timely deployment of mitigation measures, or monitoring controls. This allows organizations to proactively address vulnerabilities and reduce their attack surface (Protect), while continuous scanning provides ongoing visibility into potential risks (Detect).

Best Practices for Alignment

CSF Mapping: **This demonstrates a comprehensive alignment.**

CSF Mapping

#				
01	Understanding Cyber Risks	Identify	ID.AM (Asset Management)	Discusses identifying critical assets (data, systems, etc.)
02	Assessing Your Cybersecurity Posture	Identify	ID.RA (Risk Assessment)	Outlines risk assessment methods aligning with the CSF's approach
03	Data Center Model	Protect	PR.AC (Access Control),PR.DS (Data Security)	Provides specific guidance on access controls and data protection measures
04	Threat Intelligence	Detect	DE.CM (Continuous Monitoring)	Explains how threat intelligence feeds into ongoing monitoring for anomalies
05	Incident Response Planning	Respond	RS.RP (Response Planning)	Outlines playbook creation, aligning with the CSF's incident response
06	Remediation Strategies	Respond	RS.MI (Mitigation)	Actions to reduce risk, aligning with the CSF's mitigation and recovery aspects
07	Technology and Tools	Detect / Respond	DE.AE (Anomaly Detection) / RS.AN (Analysis)	Explains tools like SIEM and EDR, which aid in both detecting unusual activity and analyzing incidents.

Figure 23: Comprehensive Alignment of CRMP with the NIST Cybersecurity Framework (CSF)

Cyber Risk Methodology's adaptability is crucial in today's rapidly evolving threat landscape. While the NIST framework provides the strategic "what" of cybersecurity, this methodology delivers the operational "how." It emphasizes the tactics, techniques, and procedures required to address both current and future threats, ensuring its relevance stays a step ahead. This focus on execution complements the NIST framework and prevents gaps in operationalizing a robust cybersecurity defense.

Case Studies:

The NIST Cybersecurity Framework (CSF) has been widely adopted by organizations across various sectors to improve their cybersecurity risk management. While the NIST website does not currently provide detailed case studies, there are some examples of how real-world organizations have leveraged the CSF in conjunction with a risk monitoring methodology:

The Lower Colorado River Authority, a public utility in Texas, used the CSF to assess its cybersecurity posture and develop a roadmap for improvement. By mapping its existing controls to the CSF, the organization was able to identify gaps and prioritize investments to strengthen its cybersecurity capabilities.[6]

Optic Cyber Solutions, a cybersecurity consulting firm, has helped clients in the manufacturing and healthcare sectors implement the CSF. By combining the CSF with a risk monitoring approach, these organizations were able to continuously assess their security posture and make data-driven decisions to mitigate risks.[6]

The University of Pittsburgh, a leading academic institution, adopted the CSF to enhance its cybersecurity program. The university used the framework's five core functions - identify, protect, detect, respond, and recover - to develop a comprehensive risk management strategy and monitor its effectiveness over time.[6]

These examples demonstrate how organizations can leverage the CSF in conjunction with a risk monitoring methodology to assess their current cybersecurity posture, identify areas for improvement, and implement targeted measures to enhance their overall security.[1][3][6]

Cyber Risk Monitoring Self-Assessment: Building Your Digital Fortress

This assessment is designed to help you gauge your organization's current cyber risk monitoring (CRM) maturity and identify areas for improvement. By aligning with the core functions of the NIST Cybersecurity Framework, this assessment will provide you with a comprehensive understanding of your organization's preparedness to anticipate, identify, and respond to cyber threats.

Instructions:

1. Read each question carefully.
2. Answer with "Yes," "Partially," or "No" based on your organization's current practices.
3. Refer to the scoring guide at the end to interpret your results.

Identify:

1. Do you have a dedicated Cyber Risk Intelligence (CRI) team or function in place? (Yes | Partially | No)
2. Do you regularly conduct comprehensive risk assessments that consider both physical and digital threats to your data centers and other critical assets? (Yes | Partially | No)
3. Do you have a clear understanding of your organization's risk appetite and tolerance levels? (Yes | Partially | No)
4. Do you actively gather threat intelligence from both internal and external sources to inform your risk management decisions? (Yes | Partially | No)
5. Do you leverage automation and data analytics to identify patterns, anomalies, and potential threats? (Yes | Partially | No)

Protect:

1. Do you have a layered defense strategy in place that includes technical controls (e.g., firewalls, intrusion detection systems) at different levels of your organization's infrastructure? (Yes | Partially | No)

2. Do you regularly update and patch software and systems to address known vulnerabilities? (Yes | Partially | No)

3. Do you have strong access controls and identity management in place to limit unauthorized access to sensitive data and systems? (Yes | Partially | No)

4. Do you educate employees on cybersecurity best practices and phishing awareness? (Yes | Partially | No)

5. Do you maintain physical security measures to protect your data centers and other critical infrastructure? (Yes | Partially | No)

Detect:

1. Do you have continuous monitoring capabilities in place to detect anomalous activity and potential security incidents? (Yes | Partially | No)

2. Do you use security information and event management (SIEM) systems or other tools to aggregate and analyze security logs from various sources? (Yes | Partially | No)

3. Do you conduct regular security assessments, such as vulnerability scans and penetration testing? (Yes | Partially | No)

4. Do you have a defined process for detecting, analyzing, and responding to security events and incidents? (Yes | Partially | No)

5. Do you leverage threat intelligence to proactively identify and respond to emerging threats? (Yes | Partially | No)

Respond:

1. Do you have a well-defined incident response plan that outlines roles, responsibilities, and procedures for responding to security incidents? (Yes | Partially | No)

2. Do you regularly test and update your incident response plan through tabletop exercises or other simulations? (Yes | Partially | No)

3. Do you have a process for communicating security incidents to relevant stakeholders? (Yes | Partially | No)

4. Do you have the necessary resources and expertise to investigate and contain security incidents promptly? (Yes | Partially | No)

Recover:

1. Do you have a disaster recovery plan in place that addresses both physical and digital aspects of system restoration? (Yes | Partially | No)

2. Do you regularly test your disaster recovery plan to ensure its effectiveness? (Yes | Partially | No)

3. Do you maintain off-site backups of critical data to facilitate recovery in the event of a disaster? (Yes | Partially | No)

4. Do you have a plan for communicating with stakeholders during and after a disaster or cyber incident? (Yes | Partially | No)

Scoring Guide:

- Mostly "Yes" Answers: Your organization has a mature cyber risk monitoring program and is well-prepared to identify, assess, and mitigate cyber threats.

- A Mix of "Yes," "Partially," and "No" Answers: Your organization has some strengths but also areas for improvement in its cyber risk monitoring program.

- Mostly "No" or "Partially" Answers: Your organization needs to significantly improve its cyber risk monitoring capabilities to effectively manage cyber risks and protect its critical assets.

Additional Considerations:

- Review the specific areas where you answered "No" or "Partially" to identify priority areas for improvement.

- Consult the NIST Cybersecurity Framework for detailed guidance and best practices on implementing a comprehensive cyber risk management program.

Understanding ISO/IEC 27001

Achieving Global Standards: Integrating ISO/IEC 27001 with Your CRMP

Introduction:

ISO/IEC 27001 is a globally recognized international standard for establishing and managing a comprehensive Information Security Management System (ISMS). This standard outline detailed requirements for implementing, maintaining, and continually improving security processes across an organization. Crucially, it includes risk assessment and treatment processes tailored to the organization's specific needs.

While the Cyber Risk Methodology (CRM) provides the proactive, action-oriented roadmap for managing cyber risk, ISO/IEC 27001 complements this with its focus on auditable processes and globally recognized best practices. For many organizations, ISO/IEC 27001 compliance is a business necessity, not just a good idea. It can open doors to new markets, streamline client trust-building, and simplify compliance with various data protection regulations.

Additionally, to achieve this level of rigor, organizations need a proactive, actionable approach to cyber risk management. That is where a Cyber Risk Monitoring Plan (CRMP) comes into play. It provides a structured approach for identifying, analyzing, mitigating, and continuously monitoring cyber risks, safeguarding an organization's digital assets and operations.

Mapping CRMP to ISO

Mapping CRMP to ISO

ISO Clause/Section	Relevant CRMP Chapters/Sections	Brief Explanation of Alignment
A.10 (Cryptography)	Data Security, Technology Choices	Encryption best practices likely covered in these chapters. Specific guidance aligns with A.10.1, even if the word "cryptography" isn't in your CRMP.
A.11 (Physical Security)	May not have a direct CRMP counterpart	Depends on your organization's physical risk profile. If applicable, map to facility access controls sections, etc.
A.12 (Operations Security)	Threat Intel, Monitoring, Incident Response	Proactive monitoring and response support ops security. Emphasis on log analysis and anomaly detection especially relevant to A.12.3.
A.13 (Communications Sec)	Network Security chapter (if applicable)	May touch on this, but network focus is broader than A.13. Your content might exceed A.13 if it includes secure remote access, data in transit, etc.
A.14 (System Acquisition......)	Tech Choices, Vendor Risk Management	Procurement due diligence aligns with this control area. A Vendor assessment Assessment chapter would directly support A.14.2.
A.15 (Supplier Relationships)	Chapter on Third-Party Risk (if you have one)	Aligns strongly if you cover vendor assessment in detail, beyond not just stating the need to manage suppliers.
A.16 (Incident Management)	Incident Response Planning	CRMP's playbook approach helps exceed ISO's requirements – if you cover lessons learned for continuous improvement, that's A.16.2!
A.17 (Info Sec Continuity)	Disaster Recovery, Continuity Planning	A CRMP focused on minimizing downtime supports this goal. Metrics and testing emphasis would be key to map here.
A.18 (Compliance)	Could have sections on regulatory landscape	May inform on how controls comply with specific regulations. Discussing SOC 2, PCI, etc., is highly relevant.
A.5 (Info Security Policies)	Policy & Governance chapter (if you have one)	May provide templates or guidance on policy creation, exceeding ISO's basic requirement for documented policies.
A.6 (Organization of Info Sec)	Roles & Responsibilities chapter	Guidance on the definition of specific security roles and job functions not just stating the need for organizational structures, aligns with A.6.1.
A.7 (Human Resource Security)	Training & Awareness chapter	Preparing workforce for roles, aligns with A.7.2/3. If you cover hiring/termination security, that's even stronger alignment!
A.8 (Asset Management)	Asset Identification, Vulnerability Scanning	Essential for A.8.1 inventory. Your detailed risk scoring in those chapters likely exceeds ISO's basic asset classification (A.8.2).
A.9 (Access Control)	Access Control, Identity Management	Detailed guidance on tools & and methods exceeds ISO's outline. If you have multi-factor authentication MFA-specific content, that's highly relevant here.
Annex A Control Area	Relevant CRMP Sections	How CRMP Supports Compliance
Annex A Controls	(Map to various CRMP sections)	Example: A.8 (Asset Mgt) aligns with the chapter on vulnerability scanning
Clause 4 (Context)	Understanding Cyber Risks, Assessing Posture	Risk identification feeds into the ISO's understanding of stakeholders, scope, etc.
Clause 6 (Planning)	Risk Assessment, Remediation Strategies	CRMP provides the methods for the planning actions ISO mandates
Clause 9 (Evaluation)	Document & Report, Monitor	CRMP's metrics provide the data upon which ISO evaluation is based.

Figure 24: Mapping Cyber Risk Monitoring to ISO Standards

Where ISO Goes Deeper

While a robust CRMP provides a solid foundation for cyber risk management, ISO/IEC 27001 raises the bar in a few key areas:

Documentation: ISO demands detailed documentation of an organization's Information Security Management System (ISMS). This includes not just what is done to protect information, but how it has done. Policies, procedures, risk assessments, incident reports – it all needs to be meticulously recorded. Organizations may need to enhance their documentation practices to achieve full ISO alignment.

Management Review: ISO is not just about having the right practices; it mandates formal, top-down review of the cybersecurity program at regular intervals. This ensures executive-level buy-in and accountability. Reporting structures will likely need to be adapted to include a formal review process with senior management engagement.

Benefits and Challenges

While achieving ISO/IEC 27001 certification requires significant effort, the benefits are compelling:

- **Market Access:** For many organizations, ISO certification is a prerequisite for doing business in certain sectors or with clients who demand high security standards.
- **Client Trust:** It demonstrates to customers and partners a serious commitment to information security, building confidence and streamlining the trust-building process.
- **Streamlined Audits:** ISO compliance often satisfies requirements from various regulatory bodies, reducing the burden of multiple audits with overlapping requirements.

However, it is important to acknowledge the challenges:

- **Added Workload:** ISO compliance means more documentation, meetings, and potentially even new roles dedicated to managing the ISMS.

- **Potential for Bureaucracy:** ISO adherence can sometimes lead to a focus on processes over outcomes. Maintaining a practical, risk-focused mindset is key to avoid getting bogged down in paperwork, ensuring security measures remain effective. However, a CRMP that prioritizes outcomes keeps the focus on stopping breaches, not just checking boxes.

Aligning CRMP with Industry Standard

Case Studies and Examples

In 2019, <u>First American Financial Corp.</u> Suffered a major <u>data leak</u> due to poor <u>data security</u> measures and faulty website design. Although this incident was labeled a data leak instead of a breach (no hacking involved), it shows just how easily sensitive information can fall into the wrong hands.

If First American Financial Corp had a CRMP in place they would have been able to identify the weakness in their cybersecurity and website design and fixed this threat before it became a major issue and exposing approximately 880 million files.

MGM also suffered a cyber-attack in Sept 2023, this attack occurred when a hacker group was able to launch ransomware called Scattered Spider and was able to disrupt operations by disabling the online reservations system, digital room keys slot machine and website. This resulted in the loss of approx. 8 million a day in revenue but also the trust the American people in the business that their information was safe guarded.

Continuous Improvement and Adaptation

Adopting a Dynamic Approach to Cyber Risk Management through Continuous Improvement and Adaptation.

The enemy is constantly changing and so must we.

A static approach to cybersecurity is a recipe for failure. Maintaining a robust cyber risk posture demands a mindset of continuous improvement and adaptation. This section focuses on the 'how' – establishing processes and practices within your Cyber Risk Monitoring Plan (CRMP) to learn, adjust, and optimize your security posture in response to the shifting digital battlefield.

Why Continuous Improvement Matters: The Ever-Changing Threat

Cyber adversaries continually refine their tactics, techniques, and procedures (TTPs) to bypass traditional defenses. Consider these developments:

- **State-Sponsored Attacks:** Nation-states invest heavily in offensive cyber operations, developing ever more sophisticated tools that can compromise well-protected organizations.

- **Ransomware-as-a-Service:** Readily available ransomware kits lower the barrier to entry for cybercriminals, leading to an explosion in attacks aimed at businesses of all sizes.

- **Exploiting Zero-Days:** Threat actors increasingly target newly discovered vulnerabilities before software vendors release patches, leaving organizations vulnerable.

A stagnant CRMP quickly becomes obsolete in this environment. Continuous improvement ensures your defenses evolve in tandem with threats, safeguarding your critical data and operations.

Key Elements of a Continuous Improvement Framework

1. Metrics and Measurement: Peter Drucker's famous adage, "What gets measured gets managed," holds especially true in cybersecurity. Define meaningful metrics to track the effectiveness of your CRMP:
   - Vulnerability Management KPIs:
     - Time to detect new vulnerabilities.
     - Percentage of critical vulnerabilities patched or corrected within 30 days.
     - Mean time to remediate vulnerabilities based on risk severity.
   - Incident Response KPIs:

- Time to detect an incident.
- Time to containment.
- Time to full recovery.
  - ○ Awareness and Training KPIs:
    - Training completion rates.
    - Change in phishing susceptibility (simulated tests).
    - Reduction in risky user behaviors.

2. Lessons Learned and Feedback Loops:

   ○ **Incident Analysis:** After every security incident, conduct thorough root cause analysis to identify weaknesses. This is not to assign blame, but to prevent recurrence through process changes, enhanced controls, or targeted training.

   ○ **Near-Miss Reviews:** Even thwarted attacks offer insights into adversary TTPs. Examine the attempts, understand how your defenses performed, and proactively strengthen them at potential breach points.

   ○ **Stakeholder Feedback:** Regular reviews with team members, management, and other stakeholders provide valuable input on pain points, usability of the CRMP, and emerging risks they perceive.

3. Regular Re-Assessments and Reviews:

   ○ **Risk Assessment Revisits:** Schedule full-scale risk assessments at an appropriate cadence (annually, bi-annually, or more often for high-risk sectors). Changes in your business, technological advancements, or the overall threat landscape necessitate fresh evaluations.

   ○ **Framework Alignment Updates:** Re-evaluate your CRMP against NIST CSF, ISO/IEC 27001, or other relevant standards. As they evolve, your plan may need adjustments to maintain alignment and compliance.

- **CRMP Review Meetings:** Formal meetings (quarterly or semi-annually) with stakeholders ensure the plan remains relevant and receives adequate support. This is where metrics and lessons learned drive decision-making.

4. Embracing Emerging Technologies:

- **Artificial Intelligence (AI) for Enhanced Security:** AI and machine learning are transforming cyber defense. Explore AI-powered threat detection, anomaly identification, and risk prediction tools to augment your CRMP's traditional methods.

- **Automation for Efficiency:** Investigate opportunities to automate tasks like vulnerability scanning, patching, alert triage, and even elements of incident response. This frees security teams for higher-level analysis and strategic actions.

- **Cloud-Specific Adaptations:** If your CRMP includes cloud assets or services, stay abreast of cloud security advancements. Traditional on-premise tools may not suffice, and you will need to consider shared responsibility models with cloud providers.

Continuous Improvement in Action: A Case Study

Consider the evolution of phishing attacks. Early phishing emails were often riddled with errors and easily spotted. Today, they are carefully crafted, mimicking legitimate communications and targeting specific individuals. Organizations that addressed phishing solely with basic spam filters and generic user awareness found themselves increasingly vulnerable. Successful mitigation required a continuous improvement approach:

- **Advanced Detection:** Implementing email filtering with enhanced AI-powered analysis, capable of detecting subtle behavioral and linguistic cues indicative of a phish.

- **User-Submitted Reporting:** Establishing easy ways for employees to report suspicious emails, creating a crowdsourced early-warning system.

- **Targeted Simulations:** Realistic phishing simulations tailored to employees' roles, testing their susceptibility and providing focused follow-up training.

- **Metrics:** Tracking phishing click rates, susceptibility trends by department, and near-miss reporting helped measure progress and refine defenses over time.

The Benefits of Continuous Improvement

A well-structured continuous improvement program within your CRMP yields several compelling benefits:

- **Agility and Resilience:** Proactive adaptation keeps your organization ahead of attackers, minimizing the risk of successful breaches.

- **Optimized Resource Allocation:** Metrics-driven insights help ensure resources are focused on areas of highest risk and greatest impact.

- **Reduced Liability and Improved Compliance:** Demonstrating a commitment to improvement strengthens your security posture and reduces risk in the event of an incident, facilitating compliance with regulatory requirements.

- **Stakeholder Confidence:** Proactive transparency about security efforts builds trust with customers, partners, and regulators.

Potential Challenges and Strategies for Success

- **Maintaining Momentum:** Keep initial enthusiasm high by setting achievable goals, celebrating successes, and connecting improvements to tangible business outcomes.

- **The Lure of "Shiny Objects":** Balance innovation with practical execution. Avoid chasing every new technology, instead focusing on those that best address your specific risks.

- **Data Overload:** Carefully choose which metrics matter. Establish a process for gathering and analyzing data without overwhelming your team.

In cybersecurity, stasis is your greatest enemy. Without a robust continuous improvement process, your CRMP risks obsolescence, exposing your organization to vulnerabilities. By establishing the mechanisms for measurement, learning, and iteration discussed in this section, you significantly enhance your organization's ability to stay ahead of evolving cyber threats, protecting its critical assets and ensuring long-term digital resilience.

Conclusion

The journey to robust cybersecurity requires both a strategic roadmap and a detailed implementation plan. Frameworks like the NIST CSF provide the 'what', while a CRMP delivers the 'how'. And for organizations seeking global recognition of their security posture, ISO/IEC 27001 certification demonstrates commitment to information security best practices. By combining these elements, organizations can build a defense that is not only effective, but also adaptable to an ever-evolving threat landscape.

Incident Response Plans

Developing and testing a comprehensive incident response plan is essential. This plan should outline clear steps for containing and eradicating threats, restoring systems and data, and communicating with stakeholders.

Tabletop Exercises: Stress-Testing Your Defenses

A crucial step in refining incident response plans is conducting tabletop exercises. These simulated scenarios provide a safe environment to test the effectiveness of your response procedures, identify gaps in communication or decision-making, and improve overall readiness for a real incident. By bringing together key stakeholders and walking through a hypothetical cyberattack, such as a ransomware scenario, you can uncover vulnerabilities, refine communication protocols, and ensure that everyone knows their role in an emergency.

Tabletop exercises can be tailored to address a variety of threats, allowing you to focus on specific areas of concern, such as ransomware, phishing attacks, or data breaches. The goal is to create realistic scenarios that challenge your team and uncover areas where your response plans need improvement.

Defining and Integrating Risk Appetite and Tolerance

Risk appetite and tolerance are critical components of any cyber risk management strategy. Yet, they are often overlooked or not explicitly defined. This section will address this gap by explaining how organizations can determine their risk appetite and tolerance levels and integrate these concepts into their cyber risk monitoring plan.

Understanding Risk Appetite and Tolerance

- **Risk Appetite:** This is the amount and type of risk an organization is willing to accept in pursuit of its objectives. It is a strategic decision made by senior management and reflects the organization's overall risk culture.

- **Risk Tolerance:** This refers to the acceptable level of variation from the risk appetite. It is more tactical and defines the limits within which risk can be taken for specific activities or processes.

Defining Risk Appetite and Tolerance

1. **Identify Business Objectives:** Start by clearly understanding the organization's strategic goals and objectives. This will provide the context for determining the acceptable level of risk.

2. **Assess Risk Capacity:** Evaluate the organization's financial and operational capabilities to absorb losses resulting from cyber incidents. This will help define the upper limit of risk tolerance.

3. **Develop Risk Appetite Statements:** Create clear and concise statements that describe the organization's overall risk appetite. These statements should be aligned with the business objectives and risk capacity.

4. **Establish Risk Tolerance Metrics:** Define quantitative or qualitative measures for different types of cyber risks. These metrics should be linked to the risk appetite statements and specify the acceptable levels of risk variation.

Integrating Risk Appetite and Tolerance into Cyber Risk Monitoring

1. **Align Monitoring with Risk Appetite:** Ensure that the cyber risk monitoring plan is designed to identify and track risks that exceed the organization's risk appetite and tolerance levels.

2. **Prioritize Risks:** Use risk appetite and tolerance metrics to prioritize the identification, assessment, and mitigation of cyber risks.

3. **Trigger Escalation:** Establish escalation procedures based on risk tolerance thresholds. When risks approach or exceed these thresholds, appropriate actions should be taken to reduce the risk exposure.

4. **Review and Update:** Regularly review and update the risk appetite and tolerance levels as the organization's business objectives and risk environment change.

By explicitly defining and integrating risk appetite and tolerance into the cyber risk monitoring plan, organizations can make informed decisions about their cybersecurity investments, prioritize risk mitigation efforts, and ensure that their risk management strategy is aligned with their overall business objectives.

Part III

Strategic Defense Against Cyber Threats

Meet Dr. Nikki Robinson Thought Leader.

Figure 25: Dr. Nikki Robinson

Dr. Nikki Robinson is a pioneer in our field, known for her innovative work in human systems and vulnerabilities. When I met her just before she published her second book, she reminded me that our passion for cybersecurity is a calling, not a burden. Her ability to connect and share her experiences with someone like me, who often felt isolated, is a gift I will treasure for years.

This section provides organizations with the knowledge and strategies needed to combat the ever-evolving threat of commodity cyberattacks. By detailing the nature of these threats, the significance of cyber threat intelligence and operations, and emphasizing a layered defense approach, readers are guided through the complexities of modern cybersecurity challenges. The ultimate goal is to foster a comprehensive understanding of how to anticipate, identify, and mitigate commodity threats through advanced defenses, employee training, and adherence to best practices in security policies.

The Rising Threat of Commodity Attacks

The 2017 WannaCry ransomware attack, which exploited a vulnerability in older Microsoft systems and impacted hundreds of thousands of organizations globally, serves as a stark reminder of the disruptive potential of commodity cyber threats. These threats, including viruses, worms, and Trojans, may seem simple, but they have caused widespread devastation. This section will detail both common commodity threats and the methods attackers use to exploit them, providing the foundation for proactive defense strategies. While WannaCry's impact was devastating, it is just one example of the broader category of commodity cyber threats.

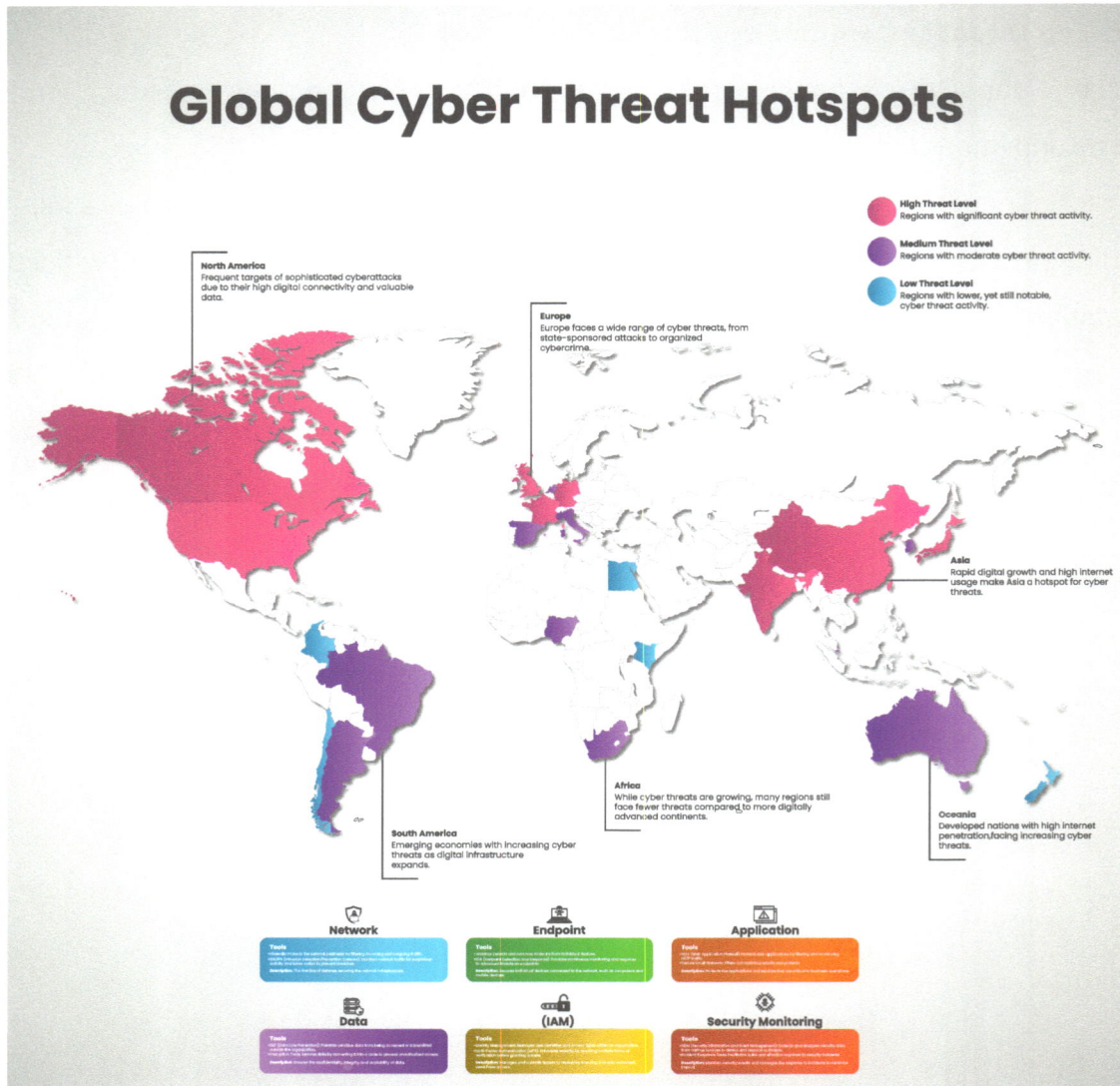

Figure 26: Global Threat Landscape: Identifying High-Risk Regions

Commodity threats often leverage automation, meaning they can spread and infect systems without direct human intervention. They typically require minimal skill to execute, allowing them to spread rapidly. Emerging commodity attacks increasingly target the growing Internet of Things (IoT) landscape, turning everyday devices into tools for cybercrime, and utilize artificial intelligence (AI) for malicious purposes like creating deepfake content for disinformation campaigns.

One particularly insidious example of a commodity threat is Emotet, a modular malware active since 2014. Primarily spread through spam emails and malicious attachments, Emotet can deliver other malware payloads, like ransomware, to compromised systems. Emotet's ability to steal sensitive data, including banking credentials and personal information, makes it a formidable threat.

To evade detection, Emotet employs techniques like polymorphism (constantly changing its code), encryption, and obfuscation. It has been used in high-profile attacks against government institutions, financial organizations, and healthcare providers. Organizations must adopt a layered approach to defense, including:

- **Advanced email filtering:** To block malicious emails at the gateway.
- **Endpoint protection:** To detect and prevent malware execution on devices.
- **User training:** To educate employees on recognizing and reporting phishing attempts.
- **Regular software updates and patches:** To address known vulnerabilities.

Understanding the attack vectors cybercriminals commonly use is crucial in defending against these threats.

Understanding the Threat Landscape

Emotet is a prime example of the dynamic nature of the threat landscape, which is continually changing, and organizations must stay vigilant to stay ahead of potential threats. Commodity threats include viruses, worms, and trojans, among others. These threats are often automated and do not require advanced skills or significant resources to execute. For example, Emotet malware is a commodity threat that has been widely used to distribute other malware, such as TrickBot and Qbot. Emotet is often spread through phishing emails and has been responsible for numerous high-profile data breaches.

Commodity threats are a significant risk to organizations, and they require a robust cybersecurity strategy to prevent, detect, and respond to them effectively Understanding the threat landscape, attack vectors, and response techniques is crucial for implementing effective cybersecurity strategies. It is crucial to have a multi-layered approach that includes technical controls, employee training, and security policies and procedures to protect against these threats.

Meet Clint Rousseau, a Thought Leader.

Figure 27: Photo of Clint Rousseau

When Clint and I first spoke about the book, his enthusiasm and specificity powered many nights of Cyber Talk. But then he said "Ransomware!"

Ransomware: A Growing Threat to Digital Assets

Ransomware attacks, a particularly insidious form of commodity threat, have emerged as a significant and evolving danger to organizations worldwide. These attacks involve malicious software that encrypts a victim's files or locks them out of their systems, demanding payment (a ransom) in exchange for restoring access. The rise in ransomware incidents, contributing to the projected cost of cybercrime exceeding $10 trillion annually by 2025, underscores the urgent need for robust mitigation strategies.

Modus Operandi: The Path to Domain Dominance

Ransomware often begins with a seemingly innocuous breach, such as compromising a low-level remote user account through a phishing email or malicious attachment. However, the initial compromise does not automatically lead to a full-blown attack. Attackers often use this initial access for reconnaissance, carefully navigating through a company's systems for weeks or even months. Their ultimate goal is to gain access to a Tier 0 (Domain Admin) credential. Without this highest-level access, a widespread ransomware attack cannot succeed.

This progression highlights the importance of a layered defense strategy. Organizations must not only protect the perimeter but also implement robust internal security controls to prevent lateral movement within their networks. A successful ransomware attack often requires compromising multiple tiers of access, from Tier 3 (laptops and workstations) to Tier 2 (servers), before ultimately reaching the critical Tier 0 (Domain Admin) level.

Emotet: A Case Study in Ransomware Delivery

Emotet, a modular malware active since 2014, exemplifies the growing sophistication of ransomware attacks. While primarily spread through spam emails and malicious attachments, its true danger lies in its ability to deliver other malware payloads, including ransomware, to compromised systems. Emotet's modular nature allows it to adapt and evolve, making it a persistent and formidable threat to organizations of all sizes.

Proactive Defense Against Ransomware

Defending against ransomware necessitates a multi-layered approach, including prevention, detection, and response strategies. However, given the critical role of Tier 0 credentials in successful attacks, organizations must *obsessively* limit, rotate, and monitor any changes to these accounts. Automated shutdowns of Tier 0 accounts upon detection of unauthorized changes can be a crucial preventative measure. Additionally, minimizing the broad use of service accounts across systems significantly reduces the risk of compromise.

Planning for Recovery: Beyond Disaster Recovery

Recovering from a ransomware attack is a complex process that extends beyond traditional disaster recovery. Organizations must implement the following:

1. **Immutable Backups:** Implement immutable backups that cannot be deleted or altered by attackers.
2. **Critical Data Isolation:** Identify and isolate critical business data in an offline, inaccessible environment.
3. **Parallel Network Recovery:** Build and test the ability to restore critical systems and data in an offline environment that is completely separate from the production network.

4. **Rapid Recovery Capability:** Be prepared to transition key operations and personnel to the offline environment within 24-48 hours.

5. **Log Retention:** Retain all relevant logs for forensic analysis to trace the attack path and identify compromised accounts.

6. **Budgeting:** Allocate a ransomware recovery budget that is 2-3 times larger than the disaster recovery budget to account for the complexity and potential for prolonged downtime associated with ransomware attacks.

By proactively addressing the ransomware threat through a combination of technical controls, employee training, vigilant monitoring of Tier 0 accounts, and a robust recovery plan, organizations can significantly reduce their risk of falling victim to these devastating attacks.

Threat Landscape

Constant Evolution

The cyber threat landscape is in a state of perpetual evolution. Attackers continuously develop new tactics, techniques, and procedures (TTPs) to exploit vulnerabilities and achieve their malicious objectives. Staying ahead of these threats requires a proactive and adaptive approach to cybersecurity.

Emerging Trends: AI and IoT in Cybercrime

Recent years have seen a rise in the use of artificial intelligence (AI) and the Internet of Things (IoT) in commodity cyberattacks. AI is being leveraged to create sophisticated phishing campaigns, generate realistic deepfake content for disinformation, and even automate the development of malware. Meanwhile, the proliferation of IoT devices has expanded the attack surface for cybercriminals, who can exploit vulnerabilities in these devices to gain access to networks and launch attacks.

Case Study: Emotet

The evolution of threats like Emotet is not just theoretical. Consider the case of this modular malware, first discovered in 2014, continues to be a major threat, adapting its tactics to evade detection and deliver a variety of malicious payloads, including ransomware. Emotet's success underscores the need for robust security measures and constant vigilance in the face of evolving threats.

Attack Vectors

Exploiting Vulnerabilities for Malicious Gain

Attack vectors describe the methods cybercriminals use to breach an organization's systems or steal data. Commodity threats often exploit known vulnerabilities, such as unpatched software, weak passwords, and misconfigured systems. These flaws are easily detectable, making robust security controls essential. The 2017 Equifax data breach, caused by an unpatched vulnerability in the Apache Struts web application framework, underscores the criticality of timely security updates.

Common Attack Vectors:

- **Phishing Attacks:** Attackers send emails impersonating trusted sources to trick recipients into revealing sensitive information or clicking malicious links.

- **Malware:** Malicious software (e.g., the Stuxnet worm) is used to gain unauthorized access or cause damage.

- **Supply Chain Attacks:** Attackers compromise elements within the supply chain (e.g., the SolarWinds attack) to infiltrate a target's network.

- **Software and System Vulnerabilities:** Outdated software and hardware provide entry points for attackers, as exemplified by the 2015 attack on Ukraine's power grid.

- **Insider Threats:** Employees or contractors with legitimate access pose a risk, as demonstrated by the 2016 Ukrainian power grid attack.

To defend against these vectors, organizations must implement strong security controls.

Prevention Strategies

Organizations need a multi-layered defense to proactively address these threats. Key strategies include:

- **Network Segmentation:** Dividing networks to limit the spread of attacks, minimizing damage even after a breach. For instance, a hospital might segment its patient data network from its administrative network to prevent a breach in one from affecting the other.

- **Access Controls:** Limiting system access based on user roles and needs can prevent unauthorized access. A strong password policy with multi-factor authentication can also help prevent breaches.

- **Intrusion Detection Systems (IDS):** IDS monitor networks for signs of malicious activity, alerting security teams to potential breaches.

- **Regular Vulnerability Assessments and Patching:** Proactively identifying and fixing security weaknesses through regular scans and prompt patching is crucial.

- **Employee Training and Awareness:** Educating staff to recognize and report phishing attacks and other suspicious activities is a first line of defense.

Detecting and responding to commodity threats is a crucial part of any cybersecurity strategy. Organizations must have effective monitoring and incident response capabilities to detect and respond to threats quickly. While proactive prevention is crucial, early detection of threats is equally important.

Detection and Response

Key to Mitigating Commodity Threats

The ability to swiftly detect and respond to commodity threats is a cornerstone of any robust cybersecurity strategy. Organizations must invest in effective monitoring and incident response capabilities to rapidly identify and address these dangers. This includes implementing Security Information and Event Management (SIEM) solutions, conducting regular security assessments, and developing a comprehensive incident response plan.

Illustrative Examples

- **Twitter Breach (2020):** A commodity threat compromised high-profile Twitter accounts. However, Twitter's detection and response capabilities minimized the potential damage.
- **SolarWinds Orion (2020):** This supply chain attack highlights the importance of timely threat detection. Microsoft's quick identification of malicious code and the "kill switch" they published helped disrupt the attack.

Key Components:

Detection:

- **SIEM tools:** These tools aggregate and analyze log data from various sources, enabling security teams to identify patterns, anomalies, and potential threats.
- **Intrusion Detection and Prevention Systems (IDPS):** IDPS monitor network traffic in real-time, detecting and potentially blocking malicious activity.
- **Endpoint Detection and Response (EDR) tools:** EDR solutions monitor endpoint devices like laptops and servers for signs of compromise, providing detailed insights into attacks and enabling swift response.

Response:

- **Incident response plans:** These plans outline clear steps, assign responsibilities, and allocate resources for effective response in case of a security incident.

- **Threat intelligence:** This helps organizations stay informed about the latest threats and attack techniques, enabling them to adapt their defenses proactively.

- **Forensic analysis:** This involves investigating the root cause of a security incident to understand how it happened and prevent similar incidents in the future.

- **Patch management:** Promptly applying security patches is essential to address known vulnerabilities and prevent further exploitation.

- **Communication and coordination:** Open communication and collaboration among all stakeholders are crucial for a swift and effective response.

By implementing a combination of prevention, detection, and response strategies, organizations can create a robust defense against commodity threats, minimizing their impact and protecting their valuable assets.

Proactive Defense: Prevention and Mitigation

Effective detection mechanisms are only the first step. In the event of a breach, a well-defined response strategy is essential to minimize damage to the business. Proactive defense against commodity threats demands a holistic approach combining technical controls, employee training, and robust security policies. Organizations must prioritize:

- **Technical Controls:** Firewalls, antivirus software, and intrusion detection systems form a critical first line of defense.
- **Employee Awareness:** Regular security training empowers employees to recognize and report potential threats.
- **Security Policies and Procedures:** Ensure consistent enforcement of security measures across the organization.

Illustrative Examples

- **Marriott Data Breach (2018):** Emphasizes the fallout of inadequate security controls and highlights the need for regular security assessments.
- **NotPetya Malware Outbreak (2017):** Underscores the criticality of timely patch management and security updates.

The Role of SIEM Systems

Security Information and Event Management (SIEM) systems play a crucial role in preventing and mitigating commodity threats, especially for critical infrastructure organizations. SIEMs aggregate and analyze log data from various network sources, identifying potential security incidents in real-time for swift response. They can also automate incident response processes.

Key Considerations for SIEM Implementation

Successful SIEM deployment requires careful planning, configuration, and regular maintenance to ensure optimal effectiveness.

Reasons for Improvement

- **Clear Structure:** Focused headings introduce key concepts.
- **Concise examples:** Reinforce the importance of proactive defense.
- **Actionable Language:** Emphasizes the practical steps organizations must take.
- **SIEM Integration:** Highlights SIEMs' role within a broader security strategy.

Cybersecurity Risk Management and Mitigation

Mitigating Third-Party Risks

To address the risks highlighted in this scenario, organizations can:

- **Conduct Vendor Security Audits:** Thoroughly assess the cloud provider's security practices, certifications, and incident response capabilities.

- **Implement Strong Contractual Agreements:** Ensure contracts with vendors include robust security requirements and data protection clauses.

- **Continuous Monitoring:** Regularly review vendor performance and security posture to identify and address any emerging risks.

Mitigation Strategies for Cybersecurity Risks

Effective cybersecurity risk management requires a multifaceted approach, integrating various strategies to address identified risks. This section outlines strategies such as mitigation, transfer, avoidance, and acceptance, with a focus on deploying effective security controls.

Mitigation: Implement layered security measures to reduce the vulnerability of systems and the potential impact of threats. Examples include:

- **Advanced Endpoint Protection:** Deploy next-generation antivirus solutions that utilize machine learning and behavior analysis to detect and block sophisticated attacks.

- **Intrusion Detection Systems (IDS):** Use IDS to monitor network traffic for suspicious activity and known threat patterns.

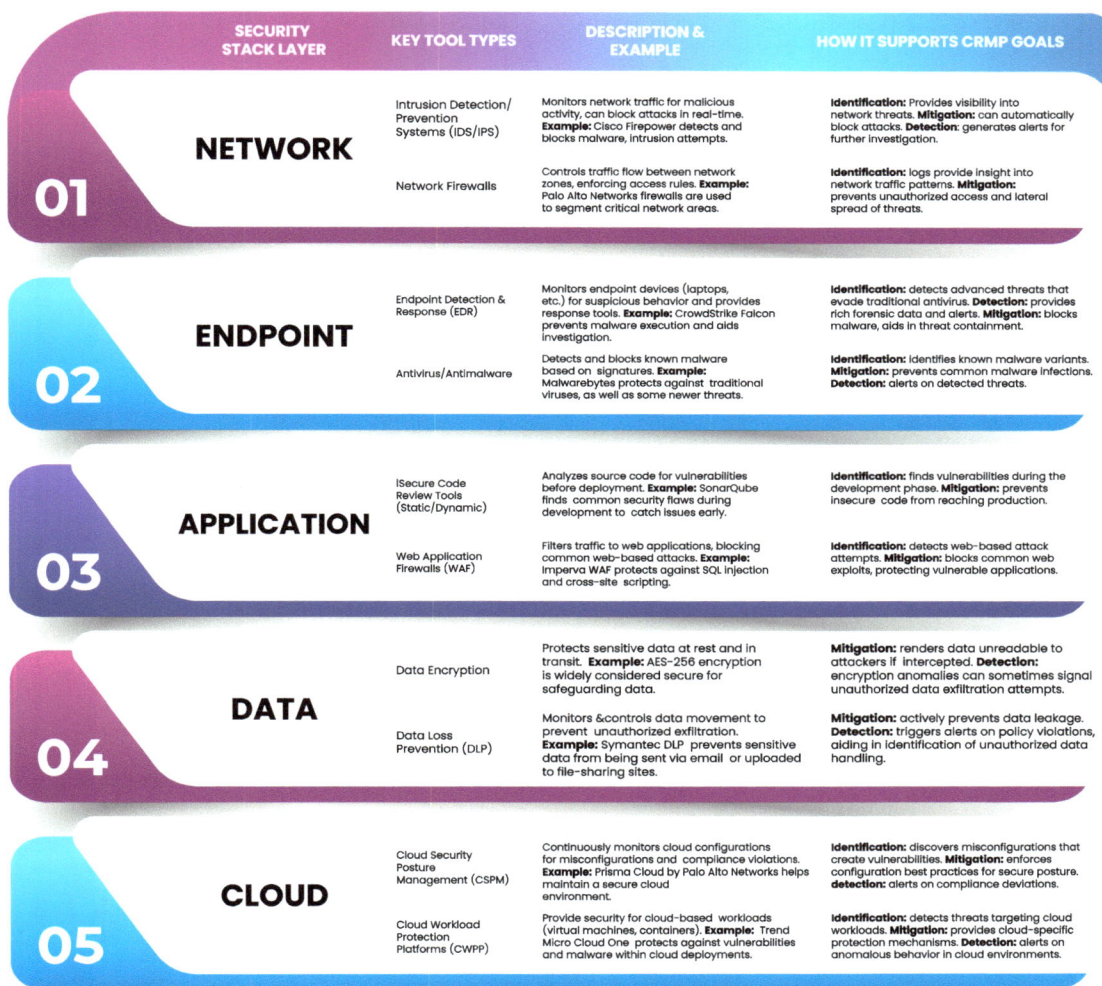

SECURITY STACK LAYER	KEY TOOL TYPES	DESCRIPTION & EXAMPLE	HOW IT SUPPORTS CRMP GOALS
01 NETWORK	Intrusion Detection/ Prevention Systems (IDS/IPS)	Monitors network traffic for malicious activity, can block attacks in real-time. **Example:** Cisco Firepower detects and blocks malware, intrusion attempts.	**Identification:** Provides visibility into network threats. **Mitigation:** can automatically block attacks. **Detection:** generates alerts for further investigation.
	Network Firewalls	Controls traffic flow between network zones, enforcing access rules. **Example:** Palo Alto Networks firewalls are used to segment critical network areas.	**Identification:** logs provide insight into network traffic patterns. **Mitigation:** prevents unauthorized access and lateral spread of threats.
02 ENDPOINT	Endpoint Detection & Response (EDR)	Monitors endpoint devices (laptops, etc.) for suspicious behavior and provides response tools. **Example:** CrowdStrike Falcon prevents malware execution and aids investigation.	**Identification:** detects advanced threats that evade traditional antivirus. **Detection:** provides rich forensic data and alerts. **Mitigation:** blocks malware, aids in threat containment.
	Antivirus/Antimalware	Detects and blocks known malware based on signatures. **Example:** Malwarebytes protects against traditional viruses, as well as some newer threats.	**Identification:** identifies known malware variants. **Mitigation:** prevents common malware infections. **Detection:** alerts on detected threats.
03 APPLICATION	iSecure Code Review Tools (Static/Dynamic)	Analyzes source code for vulnerabilities before deployment. **Example:** SonarQube finds common security flaws during development to catch issues early.	**Identification:** finds vulnerabilities during the development phase. **Mitigation:** prevents insecure code from reaching production.
	Web Application Firewalls (WAF)	Filters traffic to web applications, blocking common web-based attacks. **Example:** Imperva WAF protects against SQL injection and cross-site scripting.	**Identification:** detects web-based attack attempts. **Mitigation:** blocks common web exploits, protecting vulnerable applications.
04 DATA	Data Encryption	Protects sensitive data at rest and in transit. **Example:** AES-256 encryption is widely considered secure for safeguarding data.	**Mitigation:** renders data unreadable to attackers if intercepted. **Detection:** encryption anomalies can sometimes signal unauthorized data exfiltration attempts.
	Data Loss Prevention (DLP)	Monitors &controls data movement to prevent unauthorized exfiltration. **Example:** Symantec DLP prevents sensitive data from being sent via email or uploaded to file-sharing sites.	**Mitigation:** actively prevents data leakage. **Detection:** triggers alerts on policy violations, aiding in identification of unauthorized data handling.
05 CLOUD	Cloud Security Posture Management (CSPM)	Continuously monitors cloud configurations for misconfigurations and compliance violations. **Example:** Prisma Cloud by Palo Alto Networks helps maintain a secure cloud environment.	**Identification:** discovers misconfigurations that create vulnerabilities. **Mitigation:** enforces configuration best practices for secure posture. **detection:** alerts on compliance deviations.
	Cloud Workload Protection Platforms (CWPP)	Provide security for cloud-based workloads (virtual machines, containers). **Example:** Trend Micro Cloud One protects against vulnerabilities and malware within cloud deployments.	**Identification:** detects threats targeting cloud workloads. **Mitigation:** provides cloud-specific protection mechanisms. **Detection:** alerts on anomalous behavior in cloud environments.

Figure 28: The Security Stack: A Layered Approach to Cyber Defense

Transfer: Leverage insurance policies to transfer the financial risk associated with cybersecurity breaches. Collaborate with third-party vendors for specialized security services, ensuring they adhere to stringent security standards.

Avoidance: Refrain from engaging in activities that pose a high cybersecurity risk when possible. For example, avoiding the storage of sensitive data on systems accessible from the internet can reduce the attack surface.

Acceptance: Some risks may be accepted if they fall below the organization's risk tolerance threshold or if the cost of mitigation exceeds the potential impact. This decision should be informed by a comprehensive risk assessment.

Technology and Tools for Cyber Risk Monitoring

Effective cyber risk monitoring relies on a layered approach, often referred to as the security stack. This stack comprises tools and technologies that operate at different levels of an organization's IT infrastructure, from the network perimeter to individual endpoints and the cloud.

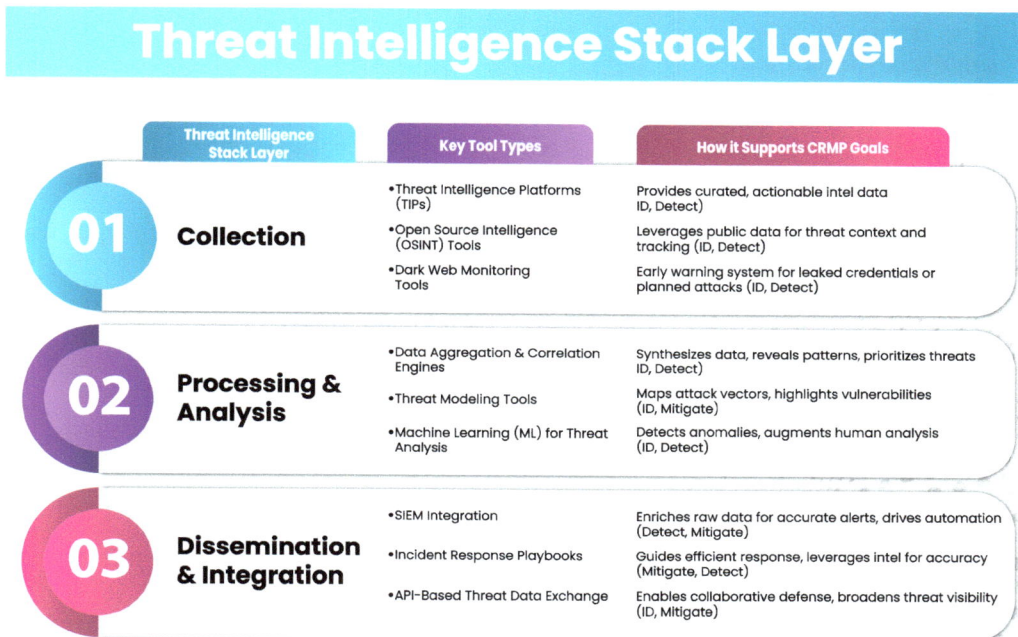

Threat Intelligence Stack Layer		Key Tool Types	How it Supports CRMP Goals
01	Collection	• Threat Intelligence Platforms (TIPs)	Provides curated, actionable intel data ID, Detect)
		• Open Source Intelligence (OSINT) Tools	Leverages public data for threat context and tracking (ID, Detect)
		• Dark Web Monitoring Tools	Early warning system for leaked credentials or planned attacks (ID, Detect)
02	Processing & Analysis	• Data Aggregation & Correlation Engines	Synthesizes data, reveals patterns, prioritizes threats ID, Detect)
		• Threat Modeling Tools	Maps attack vectors, highlights vulnerabilities (ID, Mitigate)
		• Machine Learning (ML) for Threat Analysis	Detects anomalies, augments human analysis (ID, Detect)
03	Dissemination & Integration	• SIEM Integration	Enriches raw data for accurate alerts, drives automation (Detect, Mitigate)
		• Incident Response Playbooks	Guides efficient response, leverages intel for accuracy (Mitigate, Detect)
		• API-Based Threat Data Exchange	Enables collaborative defense, broadens threat visibility (ID, Mitigate)

Figure 29: Threat Intelligence Stack: Enabling Proactive Cyber Risk Monitoring

Tool Type	Description
01 Threat Intelligence Platforms (TIPs)	Centralized hubs for collecting/managing threat data. **Example:** ThreatConnect for feed integration.
02 Open Source Intelligence (OSINT) Tools	Software to gather/analyze public info on threats. **Example:** Maltego visualizes threat connections.
03 Dark Web Monitoring Tools	Services that scan the dark web for mentions of your organization's data. **Example:** Digital Shadows
04 Data Aggregation & Correlation Engines	Collects data from multiple sources, finds patterns. **Example:** Splunk ES aggregates intel & logs.
05 Threat Modeling Tools	Helps visualize attack vectors. **Example:** Microsoft Threat Modeling Tool maps potential scenarios.
06 Machine Learning (ML) for Threat Analysis	Algorithms for detecting anomalies or subtle threatindicators. **Example:** Palo Alto Networks Cortex XDR.
07 SIEM Integration	Connects your TIP to your SIEM system. **Example:** McAfee Threat Intelligence Exchangeintegrates with SIEMs.
08 Incident Response Playbooks	Pre-defined instructions for incident response. **Example:** Crowdstrike Falcon Response provides playbooks.
09 API-Based Threat Data Exchange	Allows organizations to securely share threat intel. **Example:** STIX standard facilitates sharing.

Figure 30: Essential Tools in the Threat Intelligence Stack

Evaluate and Define Remediations

Crafting Tailored Remediation Plans

Evaluating and defining remediation plans is crucial for tailoring security measures to an organization's specific needs, ensuring optimal resource allocation and risk reduction.

- **Purpose:** The primary goal is to formulate a strategic action plan addressing the vulnerabilities identified during the risk analysis process. This involves prioritizing remediation efforts based on the severity of vulnerabilities and their potential impact on the organization.

- **Business Benefit:** By selectively implementing remediation actions, organizations can efficiently allocate resources towards the most critical vulnerabilities, enhancing their security posture while optimizing return on investment.

Operations:

- **Security Configuration Management:** Ensure that all systems are configured according to industry best practices and security standards, such as the CIS Benchmarks.

- **Vulnerability Patching:** Establish a rigorous patch management program to promptly apply security patches and updates to software and systems.

- **Security Training and Awareness:** Develop comprehensive training programs to educate employees about cybersecurity best practices, phishing awareness, and the importance of following security policies.

Additional Insights and Considerations

- **Integration of Threat Intelligence:** Incorporate real-time threat intelligence into the cybersecurity strategy to stay ahead of emerging threats and tailor remediation efforts accordingly.

- **Automated Response Mechanisms:** Utilize automated tools for rapid response to detected threats, minimizing the window of opportunity for attackers.

- **Regulatory Compliance:** Ensure that all remediation plans and cybersecurity practices align with relevant regulatory requirements, reducing legal and financial risks.

- **Continuous Improvement:** Adopt a continuous improvement approach to cybersecurity, regularly reviewing and updating the risk management strategy based on new threats, vulnerabilities, and organizational changes.

Conduct Cost-Benefit Analysis

After identifying the most relevant attack vectors for your organization, conducting a careful cost-benefit analysis is crucial for prioritizing mitigation investments. This process enables organizations to make informed decisions about implementing security measures, ensuring that cybersecurity investments are both effective and cost-efficient. While exact calculations may not always be feasible, cost-benefit analysis helps in steering clear of both under-investment and overspending on controls with minimal return on investment.

- **Purpose:** To assess the financial implications of proposed remediation actions, considering both the direct costs of implementation and the potential benefits of risk reduction.

- **Business Benefit:** A cost-benefit analysis allows organizations to make informed decisions about which cyber and physical measures to implement, ensuring that their cybersecurity investments are both effective and cost-efficient.

Operations:

- Utilize risk management frameworks, such as NIST or ISO, to conduct cost-benefit analysis of potential security investments.

- Conduct regular security awareness training for employees to minimize the risk of human error leading to a security incident.

- Apply threat modeling techniques to assess potential impacts of security incidents and prioritize investments accordingly.

Different industries often adopt various approaches to this analysis, largely due to differences in complexity and the ability to quantify benefits and costs. The key for each organization is to select a process that aligns with its structure and operations. It is also critical to continually update these processes to reflect changes in inputs and outputs, thereby enhancing precision and accuracy. Regular benchmarking is essential for maintaining relevance and effectiveness.

Analysis Types and Their Applications

Analysis Types and Their Applications

Type of Analysis	Focus	Primary Metric(s)	Complexity	Source Organization	Equations	Common Applications
Cost-Benefit Analysis (CBA)	Economic efficiency	Net Present Value (NPV), Benefit-Cost Ratio (BCR)	Moderate-High	Public agencies, private sectors, consultancies	$NPV = \Sigma$ (Benefits - Costs) / $(1+r)^t$ $BCR = \Sigma$ Benefits / Σ Costs	Projects with quantifiable financial costs &benefits
Risk-Benefit Analysis (RBA)	Balancing potential harms & benefits	Qualitative & semi-quantitative assessments of likelihood & severity	Variable	Regulatory agencies, industry, public health organizations	No universal equation: risk matrices often used	Decisions involving potential safety, health, or environmental risks
Fiscal Impact Analysis (FIA)	Budgetary impact on local governments	Net fiscal impact (revenue vs. costs)	Moderate	Local governments, economic development agencies	Net Fiscal Impact = Projected Revenues - Projected Costs	Land use, infrastructure development, economic development projects
Social Impact Analysis (SIA)	Community & societal impact	Qualitative and quantitative indicators	High	Community organizations, NGOs, government agencies, consultants	No universal equation: impact frameworks vary	Projects with social, cultural, environmental, & economic implications

Figure 31: The Analysis Types and Their Applications

Examples of Organizations That Failed to Protect Against Commodity Threats

Numerous organizations have suffered significant data breaches and other cyber incidents because of failing to adequately protect themselves against commodity threats. For example:

- In 2017, Equifax suffered a data breach that exposed the personal information of over 143 million customers. The breach was the result of a failure to patch a known vulnerability in the company's web application software, which allowed hackers to gain access to sensitive data.

- The WannaCry ransomware attack that occurred in 2017 affected hundreds of thousands of computers worldwide. The attack exploited a vulnerability in Microsoft's Windows operating system, which had been previously identified, and a

173

patch had been released. However, many organizations failed to apply the patch in a timely manner, leaving them vulnerable to the attack.

- In 2018, the Marriott data breach was caused by a commodity threat that had been present in the company's systems for several years. Marriott could have prevented the breach by implementing robust security controls and conducting regular security assessments.

Part IV

Emerging Threats and Future Trends

Building a Resilient Digital Future

"The more things change, the more they stay the same. What has been will be again, what has been done will be done again; there is nothing new under the sun." - Ecclesiastes 1:9

Two decades ago, a computer in every home was farfetched. A decade ago, cell phones in every home were ludicrous. Five years ago, working from home for millions seemed impossible. Today, it is impossible to think of a home without a device fully integrated into our way of life.

Technology transforms how we live, make choices, and interact with the world. But this rapid evolution also brings new challenges, particularly in the realm of cybersecurity. Emerging threats like AI-powered attacks, vulnerabilities in the growing Internet of Things (IoT), and the rise of disinformation campaigns are not just technical problems—they are *human* problems.

To build a resilient digital future, we must shift our focus from solely protecting data to empowering the people who use and interact with technology. Cybersecurity is no longer just the domain of IT professionals; it is a shared responsibility that involves every individual.

A People-Centered Approach to Cybersecurity

A people-centered approach recognizes that humans are both the greatest asset and the greatest vulnerability in cybersecurity. By fostering a culture of awareness, education, and vigilance, we can transform users from potential targets into active defenders against cyber threats.

This means:

- **Educating:** Providing ongoing training on the latest threats and how to recognize them.

- **Empowering:** Equipping individuals with the tools and knowledge to make informed decisions online.

- **Engaging:** Creating a culture where security is everyone's responsibility and reporting suspicious activity is encouraged.

The Human Element in the AI Age

Artificial intelligence (AI) is rapidly changing the cybersecurity landscape. While AI can be a powerful tool for both attackers and defenders, it is crucial to remember that humans remain at the heart of this battle.

AI can help us detect and respond to threats more quickly and efficiently, but it cannot replace human judgment, intuition, and ethical decision-making. We must harness AI's power responsibly, ensuring that it complements, rather than replaces, our human capacity for understanding and responding to complex threats.

Cyber Risk Monitoring: A Human-Led, Tech-Powered Approach

Cyber risk monitoring (CRM) is a critical component of a people-centered cybersecurity strategy. By proactively identifying and assessing threats, organizations can empower individuals with the information they need to make safe choices online.

This requires a multi-faceted approach:

- **Technical Controls:** Implementing robust security measures like firewalls, intrusion detection systems, and endpoint protection.

- **Threat Intelligence:** Staying informed about the latest threats and attack techniques.

- **Continuous Monitoring:** Regularly assessing vulnerabilities and adapting defenses to address emerging threats.
- **Human-Centered Training:** Educating users about the risks and empowering them to report suspicious activity.

Building a Resilient Future Together

Just as law enforcement agencies have adapted to combat new threats throughout history, we must now rise to meet the challenges of the digital age. A resilient digital future is one where technology serves humanity, not the other way around.

By embracing a people-centered approach to cybersecurity, we can create a world where individuals are empowered to protect themselves, organizations are better equipped to mitigate risks, and technology continues to improve our lives without compromising our safety and security. We built the digital world; now we must protect it, together.

Internet of Things (IoT): Expanding the Attack Surface and Amplifying Risks

The proliferation of Internet of Things (IoT) devices, encompassing everything from smart home appliances and industrial sensors to medical equipment and connected vehicles, has ushered in a new era of convenience and efficiency. However, this rapid expansion of interconnected devices has also significantly broadened the attack surface for cybercriminals, introducing new vulnerabilities and escalating risks for organizations and individuals alike.

IoT Vulnerabilities: A Target-Rich Environment

IoT devices often come with inherent vulnerabilities, including:

- **Weak or Default Credentials:** Many IoT devices ship with weak default passwords or lack robust authentication mechanisms, making them easy targets for brute-force attacks or unauthorized access.

- **Insecure Communication Protocols:** The communication protocols used by some IoT devices may lack encryption or authentication, leaving data transmissions vulnerable to interception or tampering.

- **Limited Patching Mechanisms:** Many IoT devices lack the ability to receive regular security updates or patches, leaving known vulnerabilities unaddressed and exploitable by attackers.

- **Resource Constraints:** Some IoT devices have limited processing power and memory, making them susceptible to resource exhaustion attacks or the installation of malicious software.

Attack Scenarios: A Growing Threat Landscape

These vulnerabilities make IoT devices attractive targets for a wide range of cyberattacks, including:

- **Botnet Attacks:** Hackers can compromise large numbers of IoT devices to create botnets, which can be used for various malicious purposes, such as launching distributed denial-of-service (DDoS) attacks, sending spam emails, or spreading malware.

- **Data Exfiltration:** Attackers can exploit vulnerabilities in IoT devices to gain access to sensitive data, such as personal information, financial data, or intellectual property.

- **Ransomware Attacks:** IoT devices can be targeted with ransomware, encrypting critical data, or locking users out of their devices until a ransom is paid.

- **Physical Attacks:** In critical infrastructure environments, compromised IoT devices can be used to disrupt operations or cause physical damage. For example, hackers could manipulate industrial control systems to cause a manufacturing plant to malfunction or shut down.

Mitigation Strategies: A Proactive Defense

To mitigate the risks posed by IoT devices, organizations must adopt a multi-layered approach that includes:

- **Strong Device Security:** Ensure that IoT devices are configured with strong passwords and implement multi-factor authentication wherever possible.

- **Secure Communication Protocols:** Use encryption and authentication protocols to protect data transmitted between IoT devices and other systems.

- **Regular Updates and Patching:** Stay informed about security vulnerabilities and apply updates and patches promptly to mitigate risks.

- **Network Segmentation:** Segment IoT devices from critical networks to limit the potential impact of a breach.

- **Real-Time Monitoring:** Implement continuous monitoring solutions to detect anomalies and suspicious activity on IoT devices and networks.

- **Incident Response Planning:** Develop and test incident response plans to ensure a swift and effective response to any security incidents involving IoT devices.

- **Security Awareness Training:** Educate employees and users about the risks associated with IoT devices and how to use them securely.

By understanding the vulnerabilities inherent in IoT devices and implementing proactive security measures, organizations can protect their critical assets and maintain the integrity of their operations in an increasingly interconnected world.

Artificial Intelligence: A Double-Edged Sword in Cyber Risk Monitoring

Artificial intelligence (AI), particularly machine learning (ML), is revolutionizing the cybersecurity landscape, posing both significant challenges and unprecedented opportunities for organizations striving to safeguard their digital assets.

AI-Powered Threats: A Growing Arsenal

Threat actors are increasingly leveraging AI's capabilities to launch sophisticated attacks. AI-powered phishing campaigns, for example, can craft highly personalized and convincing emails that are difficult to distinguish from legitimate communications. These malicious emails often entice unsuspecting victims to click on malicious links or divulge sensitive information, leading to data breaches and financial losses.

Moreover, AI-driven malware has emerged as a formidable threat, capable of rapidly evolving and adapting its code to evade traditional security measures. These self-learning malware strains can analyze their environment, identify vulnerabilities, and propagate autonomously, making them incredibly difficult to detect and contain.

Deepfakes, a sinister application of AI, are realistic yet fabricated media that can spread disinformation and manipulate public opinion. By creating convincingly fake videos or audio recordings, deepfakes can sow discord, undermine trust, and even influence elections, posing a serious threat to the integrity of information in the digital age.

AI-Powered Defense: A Force Multiplier

While AI presents new challenges, it also offers powerful tools for defenders. Intelligent cyber defense systems, powered by AI and ML, can analyze vast amounts of data from diverse sources, identifying patterns, anomalies, and potential threats with remarkable speed and accuracy. This enables organizations to proactively mitigate risks and respond more effectively to incidents.

AI-powered tools can automate tedious and time-consuming tasks, such as log analysis, vulnerability scanning, and threat hunting, freeing up security analysts to focus on higher-order analysis and strategic decision-making. Additionally, AI can help predict and prioritize threats, enabling organizations to allocate resources more effectively and focus on the most critical risks.

The Human-AI Partnership: The Key to Effective Cyber Risk Monitoring

The human-AI partnership is crucial for maximizing the benefits of intelligent cyber defense. AI tools are most effective when guided by human expertise and judgment. Security analysts play a vital role in interpreting AI-generated insights, understanding the context of threats, and making informed decisions based on complex and evolving circumstances. This collaboration ensures that AI-powered tools are used responsibly and ethically, with a focus on protecting both data and individual privacy.

Navigating the Future: A Continuous Evolution

As AI continues to advance, the cyber threat landscape will evolve accordingly. Organizations must stay abreast of the latest AI-powered threats and invest in AI-driven security solutions. This requires a commitment to continuous learning, adaptation, and a strategic approach that balances technological innovation with human expertise. The future of cyber risk monitoring lies in harnessing the power of AI responsibly, ethically, and collaboratively, to build a more resilient digital future.

Looking Ahead: Additional Emerging Threats

While AI is a significant factor, it is not the only emerging threat shaping the future of cyber risk monitoring. Other technologies like the Internet of Things (IoT) and quantum computing present unique challenges and opportunities.

The proliferation of IoT devices expands the attack surface for cybercriminals, introducing new vulnerabilities that require specialized security measures. Quantum computing, while still in its early stages, has the potential to disrupt current encryption standards, necessitating a proactive approach to developing quantum-resistant cryptographic solutions.

Call to Action:

The digital age is a battlefield, where cyber threats are relentless, and the stakes are high. But the power to safeguard your organization's future lies in *your* hands. Do not let your organization become another statistic in the rising tide of cybercrime.

Equip yourself and your team with the knowledge and tools necessary to build a robust defense. Develop a comprehensive cyber risk monitoring plan tailored to your organization's unique needs, one that leverages advanced technologies and military-grade intelligence to proactively identify and mitigate threats. Foster a culture of cyber awareness where every individual is empowered to be a proactive defender, ensuring not only the protection of your business but also the well-being of the families you serve.

Remember, cybersecurity is not a destination, but a continuous journey of adaptation and improvement. The threat landscape is ever evolving, and your defenses must evolve with it. Regularly reassess your risk posture, invest in emerging technologies like those offered by DBS Cyber, and remain vigilant in the face of new challenges. By embracing this mindset of continuous improvement, you will not only survive but thrive in the digital age.

Take the first step towards a more secure future today. Explore comprehensive cyber risk monitoring solutions and educational programs designed to empower individuals and organizations alike. Invest in the tools and knowledge that will protect your digital assets and ensure a brighter, safer tomorrow.

Key Influencers

Meet Dr. Nikki Robinson, Thought Leader.

Figure 32: Dr. Nikki Robinson

Dr. Nikki Robinson is a pioneer in our field, known for her innovative work in human systems and vulnerabilities. When I met her just before she published her second book, she reminded me that our passion for cybersecurity is a calling, not a burden. Her ability to connect and share her experiences with someone like me, who often felt isolated, is a gift I will treasure for years.

Dr. Robinson is a Security Architect at IBM and an Adjunct Professor at Capitol Technology University. She blends psychology and cybersecurity in her research and holds two doctorate degrees from Capitol Tech: a D.Sc. in Cybersecurity and a Ph.D. in Human Factors. Her book, "Mind the Tech Gap," published in 2022, addresses conflicts between IT and security teams from a human factors perspective.

In 2023, she won the Cyber Woman Influencer of the Year Award, highlighting her leadership in the cybersecurity community. With a background in IT Operations and Engineering, she excels in vulnerability management and security architecture, always integrating human factors into her work.

Her approachable nature and dedication to her field make her an inspiring figure. You can connect with Dr. Robinson on LinkedIn or visit her website Cyberthreads.tech.

Meet Clint Rousseau, Thought Leader.

Figure 33: Photo of Clint Rousseau

When Clint and I first spoke about the book, his enthusiasm and specificity powered many nights of Cyber Talk. But then he said "Ransomware!"

Background

- *Clint is a Senior Manager with Accenture Security bringing advanced cybersecurity and large-scale Cloud delivery leadership experience.*
- *Formerly a Microsoft Senior Architect has over 20 years of experience in Information Security design and deployment spanning across Financial Services, Industrial Manufacturing, Health Care, Education and Government.*
  - *Cybersecurity Management, Graduate Certificate University of Maryland*
  - *B.A. Information Systems, American Inter-Continental University*
  - *Certified Architect Microsoft Azure Solutions (2018)*
  - *ISC2 CISSP (2014)*
  - *Trained in Agile SCRUM delivery, Azure DevOps*
  - *Contributor, CIS Security Benchmark for Azure v1.3*

Functional Expertise

- *Epic Security and Resiliency for (Patient Health Information)*
- *Azure Cloud Cybersecurity (MITRE Attack Framework)*
- *Equinix Colo and Palo Alto Core networking design (Layer 2-3 and 7)*
- *PaaS and SaaS Security Hardening and Policy (Terraform)*
- *Azure DevSecOps design & deployment*
- *Identity-as-the-Perimeter (MFA, ADFS, Key Vault, PKI, CyberArk)*
- *Data Protection (Key Mgmt: Thales HSM FIPS 140-2 Level 3)*
- *Azure Advanced Threat Protection (ATP) Solutions*
- *MITRE Security Incident Response for Cloud (Sentinel)*
- *Automated Regulatory Compliance Reporting: PCI-DSS, ISO 27001, Azure CIS, SOC 1-2, HIPAA/HITRUST, FFIEC, NYCDFS*
- *Advanced Research POV: Securing Machine Learning / AI models and systems*
- *Security Strategy*

Meet Ryan Williams Sr., Thought Leader.

Figure 34: Photo of Ryan Williams Sr.

- **Title:** Cybersecurity Professional, Podcaster, Retired Air Force Veteran
- **Expertise:** 20+ years in network infrastructure and GRC
- **Experience:** Department of Defense, JCSE, vCISO, Consultant, RMF Assessor
- **Podcasts:** "The Other Side of the Firewall" and "Ask A CISSP"

Ryan's diverse background and hands-on experience make him a trusted voice in the cybersecurity field. His insights on the NIST Cybersecurity Framework offer a practical perspective on how this powerful tool can be implemented to strengthen an organization's defenses against evolving cyber threats.

Contribution by Ryan Williams Sr., Cybersecurity Professional, Podcaster, Retired Air Force Veteran

Greetings and salutations, good readers! My name is Ryan Williams Sr., and I am a cybersecurity professional with over 20 years of experience in network infrastructure and 10 years specializing in the Governance, Risk, and Compliance (GRC) space. As such, the NIST Cybersecurity Framework (CSF) holds a special place in my heart.

My career spans working for the Department of Defense for half my life as a network engineer and project manager. I started as a help desk technician at an NSA campus, then advanced to network technician, and ultimately became the Chief of Engineering for half of the Air Force Global network—the better half if you ask me. However, my favorite assignment was being a member of the Joint Communications Support Element (JCSE), where I had the privilege of supporting the U.S. Special Forces, a handful of three-letter agencies, Coalition Partners, and everyone in between. I helped build and secure tactical, enterprise, and bespoke networks in support of Overseas Contingency and Humanitarian Operations in Europe and Africa.

Post-military retirement, I have accumulated multiple years of experience in the private and public sectors as a vCISO, consultant, and RMF assessor. I also host 'The Other Side of the Firewall' and 'Ask A CISSP' podcasts, where we explore the latest in cybersecurity news, highlight the achievements of people of color, and interview C-Suite and Senior Leaders in the industry on how to grow a more diverse and welcoming culture. Believe it or not, across all these roles, the NIST CSF v1.1, and now v2.0, have been instrumental in enhancing the security posture of the organizations I support and providing a guide for both new assessors and maturing organizations.

How can one framework be so influential? The answer is one word: flexibility. Flexibility is not only the key to airpower—go Air Force—but also the special ingredient in your organization's cybersecurity strategy. It is not just hyperbole; for all of this amazing cybersecurity stuff to work, it has to have a solid foundation, and the NIST CSF is bedrock solid.

The NIST CSF offers numerous benefits for organizations seeking to enhance their cybersecurity risk monitoring:

- **Holistic Approach**: The framework covers all aspects of cybersecurity, ensuring a comprehensive approach to risk management.

- **Regulatory Compliance**: Adopting the NIST CSF helps organizations comply with various regulations and standards.

- **Improved Communication**: The framework fosters better communication between technical and non-technical stakeholders, facilitating a unified approach to cybersecurity.

In my personal experience, implementing the NIST CSF can significantly enhance an organization's cyber risk monitoring capabilities. Recently, while working with a mid-sized nonprofit edu-tech organization, my team and I utilized the CSF to develop a robust cybersecurity program. This included conducting thorough risk assessments, implementing multi-layered security controls, and establishing incident response protocols. As a result, the organization saw a marked improvement in overall maturity, threat detection, and response times, ultimately reducing the impact of potential cyber incidents.

In my humble opinion, the NIST CSF is one of the most valuable tools available to cybersecurity professionals today. Its structured yet flexible approach makes it accessible and practical for any organization. By fostering a proactive cybersecurity culture, the framework helps organizations not only respond to threats but also anticipate and mitigate them before they materialize.

As cyber threats continue to evolve, so too must our approaches to monitoring and managing these risks. The NIST CSF is well-positioned to adapt to these changes, with ongoing updates that incorporate emerging threats and technologies. Continuous improvement and adaptation are key to maintaining a strong cybersecurity posture, and the NIST CSF provides a solid foundation for this endeavor.

In conclusion, the NIST Cybersecurity Framework is an essential tool for effective cyber risk monitoring. Its comprehensive, flexible, and adaptable nature makes it invaluable for organizations striving to protect their assets and data from cyber threats. I encourage all cybersecurity professionals to consider adopting and leveraging the NIST CSF to enhance their risk management practices. Thank you for taking the time to read my contribution, and I look forward to further discussions on this vital topic.

CYBER RESILIENCE TTX

Treasury

Executive Overview: Department of Treasury Cyber Resilience Tabletop Exercise

Overview

The Metro Nashville Department of Treasury will conduct a cybersecurity-focused tabletop exercise (TTX) to assess and enhance its resilience against sophisticated cyber threats targeting the city's financial systems and operations. This discussion-based exercise will simulate realistic scenarios, challenging participants to make critical decisions, evaluate response procedures, and identify areas for improvement in the Treasury's cybersecurity posture.

Target Audience

This exercise is designed for Treasury department leadership, IT and cybersecurity personnel, legal counsel, communications experts, and relevant stakeholders from other city departments and external partners.

Scope

The TTX will focus on the following key areas:

- **Threat Intelligence and Risk Assessment:** Evaluating the Treasury's ability to identify, assess, and communicate potential cyber threats.
- **Incident Response and Crisis Management:** Testing the effectiveness of incident response plans, communication protocols, and decision-making processes during a cyberattack.
- **Business Continuity and Recovery:** Assessing the Treasury's ability to maintain critical financial operations and recover from a cyber incident, including financial loss mitigation and public communication strategies.
- **Cyber Insurance Evaluation:** Determining the adequacy of the city's cyber insurance coverage and its applicability to the simulated scenarios.

Objectives

The primary objectives of this exercise are to:

1. Assess and enhance the Treasury's ability to detect, contain, and recover from various cyber threats (ransomware, data breaches, phishing).

2. Strengthen communication and coordination within the Treasury, with other city departments (e.g., IT, legal), law enforcement agencies, and external partners.

3. Validate and refine the Treasury's cybersecurity plans, playbooks, and standard operating procedures.

4. Improve decision-making and leadership capabilities during a cybersecurity crisis.

5. Increase awareness and preparedness among Treasury employees and stakeholders regarding cyber threats.

6. Identify resource gaps and training needs to enhance the Treasury's cybersecurity capabilities.

7. Develop actionable recommendations to improve the Treasury's cybersecurity posture, incident response plans, communication strategies, and employee training programs.

8. Evaluate the adequacy of the city's cyber insurance coverage in relation to potential financial losses and liabilities resulting from a cyberattack.

9. **Enhance Cross-Departmental Collaboration:** Foster greater collaboration and information sharing between the Treasury and other city departments to strengthen overall cybersecurity resilience for Metro Nashville.

Purpose

This cybersecurity tabletop exercise (TTX) aims to:

1. Evaluate the effectiveness of the Treasury's incident response procedures, decision-making processes, and coordination mechanisms in various cyber crises.

2. Identify vulnerabilities within Treasury systems and their potential ramifications for the city's financial health, service delivery, and public trust.

3. Enhance communication and collaboration with key stakeholders, including other Metro Nashville departments, law enforcement agencies, regulatory bodies, financial institutions, and cybersecurity experts.

4. Strengthen the Treasury's cybersecurity posture by identifying areas for improvement in policies, incident response plans, employee training, and security controls, while aligning with financial sector standards and regulations.

5. Assess the adequacy of the city's cyber insurance coverage in relation to potential financial losses, legal liabilities, and recovery costs associated with cyberattacks.

Scenarios

ST1: Ryuk Ransomware Attack on Property Tax System

- **Threat:** The Ryuk ransomware group infiltrates the city's property tax system through a spear-phishing email, encrypting critical files and demanding a ransom in Bitcoin. They threaten to leak sensitive taxpayer data if the ransom is not paid.

- **Impact:** The attack halts property tax collection, delaying revenue for essential city services. Taxpayers cannot access records or make payments online, leading to frustration and potential financial hardships for residents. Local media reports fuel public concern and erode trust in the city's cybersecurity measures.

- **Objectives:**

  Participants will assess the Treasury's ability to:

  ○ Detect and contain the ransomware attack, including isolating infected systems and preventing lateral movement.

  ○ Evaluate response options, including data recovery from backups, engaging with law enforcement (FBI), and potentially negotiating with the attackers (with guidance from legal counsel).

  ○ Analyze the financial impact (lost revenue, recovery costs) and develop mitigation strategies.

  ○ Communicate effectively with the public and media to address concerns and maintain trust.

ST2: Data Breach Targeting Taxpayer Information

- **Threat:** A sophisticated cybercriminal group exploits a zero-day vulnerability in the Treasury's online tax filing system, exfiltrating sensitive taxpayer data (Social Security numbers, addresses, financial information). The group then demands a ransom, threatening to sell the data on the dark web if not paid.

- **Impact:** The breach exposes Nashville residents to identity theft and fraud, potentially causing widespread financial harm. Negative media coverage amplifies public outrage and damages the city's reputation.

- **Objectives:** Participants will assess the Treasury's ability to:

  ○ Detect and investigate the data breach, including identifying the attack vector and scope of data exfiltration.

- o Determine legal and regulatory obligations for notification (affected taxpayers, IRS, state comptroller) and potential liabilities.
- o Develop and execute a communication plan to inform the public, offer credit monitoring and identity theft protection services, and rebuild trust.
- o Evaluate the applicability of the city's cyber insurance policy to cover the financial and legal consequences of the breach.

ST3: Targeted Phishing Attack on Treasury Officials

- **Threat:** A spear-phishing campaign, meticulously crafted to impersonate the Tennessee Comptroller's Office, targets high-level Treasury officials. The emails contain malicious links that, when clicked, install malware designed to steal credentials and provide remote access to the Treasury's network.

- **Impact:** Attackers gain access to sensitive systems, potentially exfiltrating confidential financial data or disrupting critical operations. The breach could spread laterally within the city's network, impacting other departments.

- **Objectives:**
  Participants will assess the Treasury's ability to:
  - o Identify and thwart the phishing attack, including recognizing the sophisticated social engineering tactics used.
  - o Evaluate incident response procedures for identifying compromised accounts, containing the malware, and eradicating the threat.
  - o Assess the effectiveness of existing security controls (e.g., email filtering, multi-factor authentication) and identify areas for improvement.
  - o Develop recommendations for enhancing employee cybersecurity awareness training to prevent similar attacks in the future.

Participants

Key stakeholders, including Treasury leadership, IT and cybersecurity personnel, legal counsel, communications experts, and representatives from the FBI, DHS, the Federal Reserve, and major financial institutions (if applicable), will participate in this exercise.

Exercise Format

The TTX will follow a modular format, starting with a briefing on objectives and scenarios. Modules will cover attack detection and containment, financial impact analysis and market stabilization, state and federal government collaboration and public messaging, and forensic analysis and attribution. The exercise will conclude with a facilitated "hotwash" to discuss key takeaways and action items.

Injects

Dynamic injects, such as public outcry over delayed payments, conflicting reports on the scope of a breach, or evidence suggesting the attack's origin, will be introduced to test decision-making under pressure and to simulate real-world complexities.

Evaluation

The exercise will be evaluated on several key dimensions:

- Technical response and containment
- Leadership decision-making
- Communication and coordination with stakeholders
- Strategic considerations and identification of systemic vulnerabilities

The insights gained will inform the enhancement of cybersecurity policies, incident response plans, and communication strategies.

Key Considerations

- The TTX scenarios will highlight the unique role of the Treasury in maintaining financial stability and economic security.
- Injects will vary in complexity, ranging from technical challenges to high-level policy and ethical dilemmas.
- Lessons learned from past cyber incidents in the financial sector will be incorporated into the exercise design.

Expected Outcomes

The TTX is expected to result in:

- Enhanced readiness and resilience of the Treasury Department in the face of sophisticated cyber threats.
- Strengthened coordination and collaboration with critical partners in the public and private sectors.
- Actionable recommendations for improving cybersecurity posture and incident response plans.
- Increased awareness and understanding of the unique challenges posed by cyberattacks targeting the financial sector.
- A prioritized action plan to address identified weaknesses and vulnerabilities, along with a timeline for implementation.

By engaging in this comprehensive exercise, the Treasury Department will bolster its ability to safeguard critical financial systems and maintain public trust in the face of evolving cyber threats.

Exercise Objectives and Capabilities

Exercise Objective	Core Capability	Notes
Assess the Treasury's ability to detect and contain cyberattacks (ransomware, data breaches, phishing).	Intelligence and Information Sharing, Operational Coordination	Focus on threat intelligence sharing within the Treasury, with IT, and potentially with law enforcement.
Evaluate incident response procedures, communication, and decision-making during a cyber incident.	Operational Coordination, Planning, Public Information and Warning	Emphasize internal communication within the Treasury, as well as external communication with stakeholders (mayor's office, public, etc.).
Determine recovery strategies (system restoration, data recovery, financial mitigation).	Planning, Economic Recovery	Consider the economic impact of a cyberattack on city finances and services.
Assess the effectiveness of employee cybersecurity training and awareness programs.	Planning	This could be tied to the phishing scenario, evaluating how well employees recognize and report suspicious emails.
Evaluate communication protocols with affected taxpayers, local media, and relevant regulatory bodies (e.g., IRS).	Public Information and Warning	Focus on timely, accurate, and reassuring communication to maintain public trust.

Table 1: Exercise Objectives and Capabilities

Key Players and Roles

Metro Nashville Department of Treasury Cybersecurity Tabletop Exercise

Role	Name	Category	Description
Treasury Director		Internal	Leads strategic decision-making, coordinates with stakeholders, communicates with the public.
Chief Financial Officer		Internal	Oversees financial impact, assesses budgetary implications, develops mitigation strategies.
IT/Cybersecurity Manager		Internal	Leads the technical response, coordinates with IT staff for containment, investigation, and recovery.
Public Information Officer		Internal	Manages communication with the public, media, and stakeholders.

Treasury Staff (various roles)		Internal	Provide operational insights and contribute to decision-making.
Mayor's Office Representative		External	Provides guidance and support to the Treasury during the crisis.
Metro Nashville IT Representative		External	Assists with technical response, provides expertise on citywide IT infrastructure, coordinates with other departments.
Law Enforcement (Optional)		External	(Observer/Player) Provides guidance on legal considerations, evidence collection, and potential criminal investigations.
Cybersecurity Experts (Optional)		External	(Observer/Player) Offers specialized expertise and advice on incident response and recovery.
Lead Facilitator		Facilitator	Guides the exercise, introduces scenarios and injects, moderates discussions, ensures objectives are met.
Technical Facilitator		Facilitator	Provides technical expertise, answers questions about cybersecurity, may role-play IT staff or attackers.
Subject Matter Experts (Optional)		Facilitator	(Optional) Provide guidance on specific issues (e.g., legal counsel, communications specialists).
Observers (Optional)		External	(Optional) Representatives from other city departments or partner organizations observe to gain insights.
Role Player		External	(Acts as a player that is unavailable)

Table 2: Key Players and Roles

Internal Treasury Department

- **Treasury Director (Player):** The highest-ranking Treasury official, responsible for making strategic decisions, coordinating with external stakeholders, and communicating with the public.

- **Chief Financial Officer (Player):** Oversees the financial impact of the cyberattack, assesses budgetary implications, and works with the Director to develop mitigation strategies.

- **IT/Cybersecurity Manager (Player):** Leads the technical response to the cyberattack, coordinating with IT staff to contain the threat, investigate the incident, and implement recovery strategies.

- **Public Information Officer (Player):** Manages communication with the public, media, and other stakeholders, ensuring timely and accurate information dissemination.

- **Treasury Staff (Players):** Representatives from various divisions within the Treasury participate to provide operational insights and contribute to decision-making.

External Stakeholders

- **Mayor's Office (Player):** Represents the city's executive leadership, providing guidance and support to the Treasury during the crisis.

- **Metro Nashville IT Department (Player):** Assists with the technical response, provides expertise on citywide IT infrastructure, and helps coordinate communication with other city departments.

- **Law Enforcement (Observers or Players):** Representatives from the FBI and/or local police may participate as observers or players, providing guidance on legal considerations, evidence collection, and potential criminal investigations.

- **Cybersecurity Experts (Observers or Players):** External cybersecurity consultants or vendors may be invited to observe or participate, offering specialized expertise and advice on incident response and recovery.

- **Media (Simulated):** The exercise will simulate media inquiries and pressure to ensure the Treasury is prepared to handle public scrutiny during a cyber incident.

Exercise Facilitators

- **Lead Facilitator:** Guides the exercise, introduces scenarios and injects, moderates discussions, and ensures that the exercise objectives are met.
- **Technical Facilitator:** Provides technical expertise, answers questions about cybersecurity tools and techniques, and may role-play as IT staff or attackers.
- **Subject Matter Experts:** Additional experts (e.g., legal counsel, communications specialists) may be present to provide guidance on specific issues.

Exercise Structure

This tabletop exercise (TTX) will be a discussion-based, facilitated exercise designed to assess and enhance the Metro Nashville Department of Treasury's cybersecurity preparedness and response capabilities. Participants will engage in the following three modules:

Module One: Threat Intelligence and Risk Assessment

- **Multimedia Update:** A briefing on current cybersecurity threats relevant to municipal financial systems, highlighting potential vulnerabilities and attack vectors.
- **Discussion:** Participants will review the city's current threat intelligence sources, risk assessment methodologies, and security controls. They will discuss how to improve information sharing within the Treasury, with other city departments, and with external partners (e.g., law enforcement, financial institutions).
- **See Scenario Discussion Questions (1A)**

Module Two: Incident Response and Crisis Management

- **Multimedia Update:** A simulated cyberattack unfolds, presenting the Treasury with a specific scenario (e.g., ransomware, data breach, phishing).

- **Discussion:** Participants will work through the incident response process, focusing on detection, containment, eradication, and recovery. They will discuss communication protocols with internal and external stakeholders, decision-making under pressure, and the activation of relevant cybersecurity plans and procedures.

Module Three: Business Continuity and Recovery

- **Multimedia Update:** The cyberattack's impact on Treasury operations is assessed, highlighting financial losses, service disruptions, and potential reputational damage.

- **Discussion:** Participants will evaluate the effectiveness of the Treasury's business continuity and disaster recovery plans. They will discuss strategies for restoring critical systems and services, mitigating financial losses, and communicating with the public to rebuild trust.

Exercise Guidelines

- **Open, No-Fault Environment:** This exercise is designed to be a collaborative learning experience. All participants are encouraged to share their perspectives, ask questions, and challenge assumptions.

- **Use Existing Knowledge and Resources:** Respond to the scenarios based on your current understanding of plans, procedures, and available resources.

- **Focus on Problem-Solving:** The goal is to identify areas for improvement and develop actionable recommendations, not to assign blame or dwell on shortcomings.

- **Plausible Scenario:** While the scenario is fictional, it is designed to be realistic and relevant to the threats faced by the Metro Nashville Department of Treasury.

- **Equal Access to Information:** All participants will receive the same information simultaneously to ensure a fair and equitable exercise.

Exercise Guideline: Confidentiality

Participants are expected to maintain the confidentiality of any sensitive information, discussions, or hypothetical scenarios presented during this tabletop exercise. This includes, but is not limited to, details about the Treasury Department's systems, security vulnerabilities, incident response plans, and any personal or financial information discussed in the context of the exercise.

This information is intended solely for the purpose of improving the city's cybersecurity preparedness and should not be shared outside of this exercise without explicit authorization from the Treasury Director or designated representative.

Exercise Evaluation

The exercise will be evaluated based on the achievement of the stated objectives and the effectiveness of core capabilities demonstrated by participants. Feedback will be gathered through:

- **Participant Feedback Forms:** Participants will complete forms to assess the exercise's value, realism, and relevance to their roles.

- **Facilitator Observations:** Facilitators will observe discussions, decision-making processes, and overall engagement to identify strengths and weaknesses.

- **Evaluator Notes:** Designated evaluators will take detailed notes throughout the exercise to capture key insights and areas for improvement.

This feedback will be compiled into a comprehensive After-Action Report/Improvement Plan (AAR/IP), which will include specific recommendations for enhancing the Treasury's cybersecurity posture and resilience.

Exercise Agenda

Start Time	End Time	Activity
7:45 a.m.	8:30 a.m.	Registration
8:30 a.m.	8:45 a.m.	Welcome and Participant Briefing
8:45 a.m.	9:45 a.m.	Module One:
9:45 a.m.	9:55 a.m.	Break
9:55 a.m.	10:55 a.m.	Module Two:
10:55 a.m.	11:05 a.m	Break
11:05 a.m.	12:05 p.m.	Module Three:
12:05 p.m.	12:30 p.m.	Hot Wash

Table 3: Schedule of Events

Scenario Briefing

Metro Nashville Department of Treasury Cybersecurity Tabletop Exercise

Date:

Location:

Introduction

Welcome to the Metro Nashville Department of Treasury Cybersecurity Tabletop Exercise. This exercise aims to simulate realistic cyber threats targeting the city's financial systems and operations. By working through these scenarios, we will assess our ability to detect, respond to, and recover from cyberattacks, ultimately strengthening our overall cybersecurity posture.

Scenario 1: Ransomware Attack on Property Tax System

- **Threat:** A sophisticated ransomware group has infiltrated the city's property tax system, encrypting critical files and demanding a ransom in Bitcoin for their release.
- **Impact:** The attack disrupts property tax collection, delaying revenue for essential city services. Taxpayers are unable to access their records or make payments online. Local media outlets begin reporting on the attack, causing public concern and potential panic.
- **Objectives:**
  - Assess the Treasury's ability to detect and contain the ransomware attack.
  - Evaluate incident response procedures, including communication with IT, law enforcement (FBI, potentially local police), and the public.
  - Determine the feasibility of restoring systems from backups or negotiating with the attackers (with guidance from law enforcement).
  - Analyze the financial impact of the attack and develop mitigation strategies.

Scenario 2: Data Breach Targeting Taxpayer Information

- **Threat:** A cybercriminal group has exploited a vulnerability in the Treasury's online tax filing system, stealing sensitive taxpayer data, including Social Security numbers, addresses, and financial information.

- **Impact:** The breach exposes Nashville residents to identity theft and fraud. Local media reports on the incident, leading to public outrage and a loss of trust in the city's ability to protect sensitive information.

- **Objectives:**
  - Assess the Treasury's ability to detect and investigate the data breach.
  - Evaluate communication protocols with affected taxpayers, local media, and relevant regulatory bodies (e.g., IRS).
  - Determine legal obligations and potential liabilities for the city.
  - Develop plans for credit monitoring and identity theft protection services for affected taxpayers.

Scenario 3: Targeted Phishing Attack on Treasury Employees

- **Threat:** A highly targeted phishing campaign impersonates a trusted vendor, tricking Treasury employees into revealing login credentials or clicking on malicious links.

- **Impact:** The attackers gain access to internal systems, potentially stealing financial data or disrupting operations. The attack could also spread laterally within the city's network, affecting other departments.

- **Objectives:**
  - Assess the effectiveness of the Treasury's employee cybersecurity training and awareness programs.
  - Evaluate incident response procedures for identifying compromised accounts and containing the attack.

○ Analyze the potential scope of the attack and its impact on Treasury operations.

○ Develop recommendations for improving employee training and security controls.

Open Survey and systems agnostic description:

System/ Technology	Description	Priority of Recovery	Potential Cyberattack Impacts	Relevance to Exercise Scenarios
Banking Relationship Management	Handles payments, credit/debit processing, armored car services.		Unauthorized transactions, fraudulent payments, theft of funds, disruption of cash flow.	Ransomware (Scenario 1), Data Breach (Scenario 2), Phishing (Scenario 3)
Electronic Payment Systems	Online portals for tax payments and other financial transactions.		Interruption of services, theft of payment information, unauthorized access to taxpayer data, reputational damage.	All scenarios
Remote Deposit Capture	Allows for electronic deposit of checks and other financial instruments.		Unauthorized deposits, manipulation of financial records, theft of funds.	Ransomware (Scenario 1), Data Breach (Scenario 2)
Debt Management	Tracks and manages the city's debt obligations.		Unauthorized changes to debt records, misappropriation of funds, disruption of financial reporting.	Data Breach (Scenario 2), Phishing (Scenario 3)
Investment Management	Oversees the city's investment portfolio and financial assets.		Unauthorized trades, manipulation of investment data, financial losses.	Data Breach (Scenario 2), Phishing (Scenario 3)

System/ Technology	Description	Priority of Recovery	Potential Cyberattack Impacts	Relevance to Exercise Scenarios
Core IT Infrastructure	Servers, workstations, networking equipment, backup systems, malware detection.		Ransomware encryption, data exfiltration, system downtime, service disruption, lateral movement to other city systems.	All scenarios
Wireless Network	Provides wireless connectivity for Treasury staff and potentially public access points.		Unauthorized access points, interception of sensitive data transmitted over the network, disruption of services.	Data Breach (Scenario 2), Phishing (Scenario 3)
Cloud Services	Used for data backup, storage, and potentially other applications.		Unauthorized access to cloud data, data deletion or corruption, service disruption, ransomware attacks targeting cloud infrastructure.	All scenarios

Instructions

1. **System Identification:** Review the table of Treasury systems and technologies. Ensure all relevant systems are listed and understood by participants.

2. **Impact Assessment:** For each system, discuss and document the potential impacts of a cyberattack, considering the following:

   ○ Financial Loss: Direct financial losses (e.g., theft of funds, fraudulent transactions) and indirect losses (e.g., lost revenue, recovery costs).

   ○ Operational Disruption: Impact on essential services (e.g., tax collection, payments), employee productivity, and public access to services.

   ○ Reputational Damage: Loss of public trust, negative media coverage, impact on the city's image.

○ Legal/Regulatory Consequences: Potential fines, lawsuits, or regulatory violations.

3. **Recovery Time Objectives (RTOs):** For each system, estimate the maximum acceptable downtime before significant harm occurs. This will help prioritize recovery efforts.

4. **Prioritization:** Based on the impact assessment and RTOs, collectively rank the systems in order of priority for protection and recovery. Consider using a simple 1-5 scale (1 = highest priority, 5 = lowest priority) or another method that works for the group.

5. **Documentation and Visualization:** Create a visual representation of the prioritized systems and their interdependencies. This could be a simple list, a diagram, or a workflow chart. (See suggested layouts below)

6. **System Dependencies:** After prioritizing individual systems, have participants map out the dependencies between them. This will help identify potential cascading failures and prioritize recovery efforts for the most interconnected systems.

7. **Risk Assessment Matrix:** Introduce a risk assessment matrix (likelihood vs. impact) to add another layer of analysis to the prioritization process. This can help identify systems that may not have the highest impact individually but pose a significant risk due to their vulnerability or interconnectedness.

8. **Resource Allocation:** Have participants consider resource allocation (e.g., budget, personnel, time) when prioritizing systems for protection and recovery. This will add a practical dimension to the exercise and ensure that recommendations are feasible and sustainable.

Elevated Threat Alert and Cyberattack Concerns for Metro Nashville Treasury Department

Date:

Background: The Department of Homeland Security (DHS), in coordination with federal entities, has issued an "Elevated" Threat Alert through the National Terrorism Advisory System (NTAS), warning of a credible terrorist threat against the United States. While the alert primarily focuses on domestic anti-government terrorist groups targeting large facilities, the heightened risk environment raises concerns about potential cyberattacks on critical infrastructure, including municipal financial systems.

The Tennessee Fusion Center has relayed this alert to the Metro Nashville Treasury Department, emphasizing the need for increased vigilance and preparedness for potential cyber threats that could exploit the heightened security situation.

Discussion Questions:

1. **Threat Intelligence and Risk Assessment:**
   - How does the Treasury Department typically receive and assess information about potential threats, including NTAS alerts and cyber threat intelligence?
   - What steps would the Treasury take to evaluate the credibility and relevance of this specific threat alert to its operations and systems?
   - How does the Treasury communicate threat information internally and with external partners, such as the Metro Nashville IT Department, law enforcement, and financial institutions?
   - What additional threat intelligence sources or analysis tools could enhance the Treasury's ability to identify and assess potential cyber threats?

2. **Incident Response and Preparedness:**
   - What specific cybersecurity measures would the Treasury implement in response to this elevated threat alert?
   - How would the Treasury adapt its incident response plans and procedures to address the potential for both physical and cyber threats?
   - Does the Treasury have established communication channels with law enforcement agencies (e.g., FBI, local police) to report suspicious activity or potential cyberattacks?
   - What roles and responsibilities are assigned to Treasury staff during a cybersecurity incident, and how can these roles be clarified or strengthened?

3. **Business Continuity and Public Communication:**
   - How would the Treasury ensure the continuity of critical financial operations in the event of a cyberattack during this heightened threat environment?
   - Does the Treasury have a crisis communication plan in place to address public concerns and maintain trust in the event of a cyber incident?
   - What specific messages would the Treasury communicate to the public, employees, and other stakeholders in response to the elevated threat alert and potential cyber threats?

4. **Additional Considerations:**
   - Does the Treasury Department have access to the Homeland Security Information Network – Critical Infrastructure (HSIN-CI) portal for sharing threat information? If not, how could access be obtained?
   - What training or resources would be beneficial for Treasury staff to enhance their awareness and preparedness for cyber threats in this heightened risk environment?

    ○ Are there any specific legal or regulatory considerations that the Treasury needs to address regarding information sharing or incident response during a national security threat?

Transition from Module 1 to Module 2

Facilitator Announcement: "While the Treasury Department is still grappling with the ransomware attack on the property tax system, a new threat emerges. Initial forensic analysis of the ransomware incident reveals an alarming discovery – signs of unauthorized access and potential exfiltration of sensitive taxpayer data during the attack."

Multimedia Update:

- A slide presentation displays a timeline of events, highlighting the initial ransomware infection, followed by indicators of data exfiltration (e.g., unusual network traffic patterns, access to sensitive databases).
- A mock news article reports on the possibility of a data breach affecting Nashville taxpayers, citing unnamed sources within the Treasury Department.

Discussion Questions:

1. **Data Breach Assessment and Containment:**
   - ○ How does the Treasury Department assess the scope and severity of the potential data breach? What specific types of taxpayer data might have been compromised?
   - ○ What immediate steps should the Treasury take to contain the breach and prevent further unauthorized access to sensitive data?
   - ○ How does this new development affect the ongoing response to the ransomware attack? Should resources be reallocated, and if so, how?

2. **Legal and Regulatory Considerations:**

   o What legal obligations does the Treasury have to notify affected taxpayers, regulatory agencies (e.g., IRS), and potentially law enforcement about the breach?

   o What are the potential legal liabilities for the city in the event of a confirmed data breach?

   o Are there any specific state or local regulations that govern the handling of taxpayer data breaches?

3. **Public Communication and Reputation Management:**

   o How should the Treasury communicate this new development to the public, considering the ongoing concern about the ransomware attack?

   o What steps can the Treasury take to mitigate potential reputational damage and maintain public trust?

   o What resources or services should the Treasury offer to affected taxpayers (e.g., credit monitoring, identity theft protection)?

4. **Long-Term Implications:**

   o How will this incident affect the Treasury's cybersecurity priorities and future investments?

   o What lessons can be learned from this combined ransomware and data breach incident to strengthen the city's overall cybersecurity posture?

   o How can the Treasury better prepare for and respond to similar incidents in the future?

Facilitator Note: Encourage participants to consider the interconnectedness of these two incidents and the potential cascading effects of a cyberattack. Emphasize the importance of adaptability, flexibility, and effective communication in managing a complex and evolving crisis.

Transition from Module 2 to Module 3

Facilitator Announcement: "While the Treasury Department continues to address the fallout from the data breach, another incident is unfolding. Treasury employees are reporting an influx of suspicious emails that appear to be from a trusted vendor. This raises concerns about a potential targeted phishing attack aimed at exploiting the current heightened security environment."

Multimedia Update:

- A slide displays examples of the phishing emails, highlighting their deceptive tactics (e.g., urgent language, requests for sensitive information, links to malicious websites).

- A mock internal memo from the IT department warns employees about the phishing campaign and reminds them of security best practices.

Discussion Questions:

1. **Phishing Attack Identification and Containment:**
   - How does the Treasury Department identify and verify the legitimacy of phishing emails? What tools or processes are in place to detect and block these threats?
   - What steps should the Treasury take to immediately contain the phishing campaign and prevent employees from falling victim to the attack?
   - How can the Treasury leverage existing security controls (e.g., email filtering, multi-factor authentication) to mitigate the risk of successful phishing attacks?

2. **Employee Awareness and Training:**
   o How effective are the Treasury's current cybersecurity awareness and training programs in preparing employees to recognize and report phishing attempts?
   o What additional training or resources would be beneficial to reinforce security best practices and raise awareness about the evolving tactics of phishing attacks?
   o How can the Treasury create a culture of cybersecurity vigilance among employees, encouraging them to actively participate in protecting the department's systems and data?

3. **Incident Response and Communication:**
   o If employees have already fallen victim to the phishing attack, what incident response procedures should be initiated?
   o How should the Treasury communicate the phishing campaign to employees, providing clear guidance on how to identify and report suspicious emails?
   o Should the Treasury consider temporarily restricting access to certain systems or data while the phishing attack is being investigated and contained?

4. **Long-Term Resilience:**
   o How can the Treasury improve its overall resilience to phishing attacks and other social engineering tactics?
   o What additional security controls or technologies could be implemented to reduce the risk of successful phishing attacks in the future?
   o How can the Treasury foster a collaborative relationship with IT and cybersecurity staff to ensure a coordinated and effective response to evolving threats?

Facilitator Note: Emphasize the importance of employee awareness and training in preventing phishing attacks. Encourage participants to consider the psychological aspects of phishing and how attackers exploit trust and urgency to manipulate victims. Highlight the need for a multi-layered approach to cybersecurity, combining technical controls with ongoing education and awareness programs.

Evaluation Plan:

Metro Nashville Department of Treasury Cybersecurity Tabletop Exercise

The evaluation of this tabletop exercise will focus on assessing the Treasury's performance in relation to the stated exercise objectives and core capabilities. A multi-pronged approach will be used to gather feedback and observations throughout the exercise:

1. Participant Feedback Forms

At the conclusion of each module, participants will complete a brief feedback form assessing:

- Relevance of the scenario to their roles and responsibilities.
- Clarity of objectives and instructions.
- Effectiveness of the discussions and decision-making processes.
- Overall value of the exercise for their professional development.

2. Facilitator Observations and Evaluation Criteria

Throughout the exercise, facilitators will observe and evaluate participant interactions and decision-making using the following criteria and rating scale:

Criterion	Description	Not Effective (1)	Needs Improvement (2)	Effective (3)	Highly Effective (4)
Threat Intelligence & Risk Assessment	Ability to gather, analyze, and share relevant threat intelligence, assess risks, and make informed decisions based on available information.				
Incident Detection & Analysis	Timeliness and accuracy of identifying and analyzing potential cyber threats, including recognizing indicators of compromise and understanding the potential impact.				
Incident Response & Containment	Effectiveness of implementing incident response procedures, containing the attack, and minimizing damage to systems and data.				
Decision-Making & Leadership	Ability of leadership to make timely and informed decisions under pressure, effectively communicate those decisions, and provide clear guidance to staff.				
Communication & Coordination	Quality and timeliness of communication both within the Treasury department and with external stakeholders (e.g., Metro IT, law enforcement, public). Ability to coordinate actions effectively across different teams and organizations.				

Business Continuity & Recovery	Effectiveness of activating and executing business continuity and disaster recovery plans. Ability to restore critical systems and services and mitigate financial losses.					
Public Messaging	Clarity, accuracy, and timeliness of public communication during a cyber incident. Ability to manage public perception, address concerns, and maintain trust.					
Cyber Insurance Evaluation	Understanding of the city's cyber insurance policy coverage, terms, and limitations. Ability to assess the policy's applicability to the simulated scenarios and potential gaps in coverage.					

3. Evaluator Notes

Designated evaluators will take detailed notes throughout the exercise, focusing on:

- Key decisions made by participants.
- Challenges or bottlenecks encountered.
- Innovative solutions or strategies proposed.
- Areas where existing plans or procedures were effective.
- Gaps in knowledge, resources, or capabilities.

4. After-Action Report/Improvement Plan (AAR/IP)

The evaluation data will be compiled into a comprehensive AAR/IP, which will include:

- Summary of the exercise's strengths and weaknesses.

- Specific recommendations for improvement, prioritized based on their potential impact and feasibility.

- Action items and timelines for implementing the recommendations.

By combining quantitative and qualitative evaluation data, this plan aims to provide a comprehensive assessment of the Treasury's cybersecurity capabilities and identify actionable steps to enhance its preparedness and resilience against cyber threats.

Glossary of Key Terms

Advanced Persistent Threat (APT): A sophisticated, stealthy, and long-term cyberattack typically orchestrated by a nation-state or well-resourced group to steal information or disrupt operations.

Attack Vector: A method or pathway used by cybercriminals to gain unauthorized access to a system or network.

Business Impact Analysis (BIA): A process to determine the potential impact of a disruption to critical business operations, used to inform risk management decisions.

Commodity Threat: A common, widespread cyber threat that often relies on automation and requires minimal skill to execute, such as viruses, worms, or Trojans.

Cyber Kill Chain: A model that describes the stages of a cyberattack, from initial reconnaissance to data exfiltration.

Cyber Risk Intelligence (CRI): The process of collecting, analyzing, and interpreting information about cyber threats and vulnerabilities to inform decision-making and risk mitigation strategies. *See also: Threat Intelligence.*

Cyber Risk Monitoring Plan (CRMP): A comprehensive plan that outlines an organization's approach to identifying, assessing, and mitigating cyber risks.

Cybersecurity Intelligence: The specific subset of intelligence focused on delivering accurate and timely information about cyber threats, vulnerabilities, and the broader technological landscape.

Data Center: A physical facility that houses computer systems, storage, networking equipment, and other critical infrastructure for the purpose of storing, processing, and distributing data and applications.

Deepfake: A type of media that is artificially generated or manipulated using artificial intelligence to appear authentic but is actually fabricated.

Endpoint Detection and Response (EDR): A security solution that monitors endpoint devices (e.g., laptops, servers) for signs of compromise and provides tools for responding to security incidents.

Information Operations (IO): Actions taken to influence the information environment, including both defensive measures (e.g., protecting sensitive information) and offensive tactics (e.g., disinformation campaigns).

Indicator of Compromise (IOC): A clue or piece of evidence that suggests a system or network has been breached or compromised by a cyberattack.

Intelligence Cycle: A structured process for collecting, analyzing, and disseminating intelligence, consisting of five stages: Planning and Direction, Collection, Processing and Exploitation, Analysis and Production, and Dissemination.

Intelligence Requirements (IRs): The specific information needs of decision-makers, which guide the intelligence collection and analysis process.

Internet of Things (IoT): A network of interconnected physical devices, vehicles, home appliances, and other items embedded with electronics, software, sensors, actuators, and connectivity, enabling these objects to connect and exchange data.

Machine Learning (ML): A subset of artificial intelligence that involves algorithms and statistical models that enable computer systems to learn from data and improve their performance over time.

Mitigation: A risk management strategy that involves taking steps to reduce the probability or impact of a potential cyber risk.

Phishing: A type of social engineering attack often used to steal user data, including login credentials and credit card numbers. It occurs when an attacker, masquerading as a trusted entity, dupes a victim into opening an email, instant message, or text message.

Polymorphism: The ability of malware to constantly change its code to evade detection by antivirus software.

Ransomware: A type of malware that encrypts a victim's files and demands payment (a ransom) in exchange for the decryption key.

Security Information and Event Management (SIEM): A software solution that aggregates and analyzes log data from various sources across an organization's network to identify and respond to security threats.

Security Orchestration, Automation, and Response (SOAR): A technology that combines security tools and processes with incident response workflows to automate and streamline security operations.

Supply Chain Attack: A cyberattack that targets an organization by exploiting vulnerabilities in its supply chain, such as those in third-party vendors or software.

Threat Actor: An individual or group responsible for cyberattacks, often motivated by financial gain, espionage, or political objectives.

Threat Intelligence: Information about potential or current cyber threats, including their tactics, techniques, and procedures (TTPs). *See also: Cyber Risk Intelligence.*

Vulnerability Scanning: The process of identifying weaknesses or vulnerabilities in a system or network that could be exploited by attackers.

Zero Trust Model: A security framework that assumes no user or device is inherently trustworthy and requires strict verification of identities and permissions before granting access to resources.

Bibliography

1. Statista. "U.S. consumers and cyber crime - Statistics & Facts." Retrieved from on December 19, 2023.

2. Internet Crime Complaint Center (IC3). "2022 INTERNET CRIME REPORT." Retrieved from https://www.ic3.gov/Media/PDF/AnnualReport/2022_IC3Report.pdf on March 14, 2023.

3. Statista. "Infographic: Cybercrime Expected To Skyrocket in Coming Years." Retrieved

4. Comparitech. "300+ Terrifying Cybercrime & Cybersecurity Statistics (2024)." Retrieved from on January 19, 2024.

5. Smith, J. (2022). "The Rising Tide of Cyber Risks: A Global Perspective." Journal of Cybersecurity, 15(2), 45-62.

6. Cybersecurity Institute. (2023). "Cyber Threats and Risk Management in the Digital Age." Retrieved from https://www.cybersecurityinstitute.org/research/cyber-threats-risk-management on January 10, 2023.

7. Johnson, A., & Lee, C. (2024). "Effective Strategies for Cyber Risk Monitoring in Organizations." International Journal of Information Security, 28(4), 112-129.

8. Chen, L., & Wang, Q. (2023). "Cyber Risk Landscape: Threats, Vulnerabilities, and Impacts." Journal of Cybersecurity Management, 10(3), 87-104.

9. Cybersecurity Research Institute. (2024). "Emerging Trends in Cyber Threats and Risk Management." Retrieved from on February 5, 2024.

10. Martinez, S., & Johnson, M. (2022). "Insider Threats in Organizations: Detection and Mitigation Strategies." International Journal of Information Security, 25(1), 56-73.

11. Smith, A., & Brown, C. (2023). "Cyber Threat Landscape: Insights and Strategies." Journal of Cybersecurity, 12(4), 76-91.

12. Cybersecurity Institute. (2024). "Effective Implementation of Cyber Risk Monitoring Plans." Retrieved from https://www.cybersecurityinstitute.org/research/cyber-risk-monitoring-plans on February 15, 2024.

13. Johnson, T., & Lee, D. (2022). "Continuous Improvement in Cyber Risk Management: Lessons Learned from Case Studies." International Journal of Information Security, 27(2), 45-62.

14. Veeam. (2022). The Importance of Offsite Backups for Disaster Recovery.

15. NIST. (2016). Guide for Conducting Risk Assessments. NIST Special Publication 800-30 Rev. 1.

16. Gartner. (2021). Disaster Recovery Planning: A Guide to Successful Implementation. https://www.gartner.com/en/information-technology/insights/disaster-recovery-planning

17. [1] https://www.cybersaint.io/blog/nist-cybersecurity-framework-assessment-tool

18. [2] https://www.nist.gov/cyberframework

19. [3] https://www.telos.com/offerings/xacta-nist-csf-automation/

20. [4] https://csf.tools

21. [5] https://csrc.nist.gov/projects/cybersecurity-framework/filters

22. [6] https://www.nist.gov/cyberframework/success-stories

23. https://westoahu.hawaii.edu/cyber/global-weekly-exec-summary/alphv-hackers-reveal-details-of-mgm-cyber-attack/#:~:text=The%20hacker%20groups%20used%20social,%2C%20slot%20machines%2C%20and%20websites.

24. https://www.upguard.com/blog/biggest-data-breaches-us

Citations

- Smith, J. (2022) emphasizes the critical need for robust cyber risk monitoring in today's digital landscape to combat the evolving cyber threats and challenges organizations face.
- The Cybersecurity Institute (2023) highlights the dire consequences of cyber incidents like the SolarWinds cyberattack and stresses the importance of proactive cyber risk management to mitigate financial and reputational damages.
- Johnson, A., & Lee, C. (2024) underscore the significance of developing a comprehensive Cyber Risk Monitoring Plan (CRMP) to enhance cybersecurity posture and protect organizations from a wide range of cyber risks.
- Chen, L., & Wang, Q. (2023) emphasize the importance of understanding the broad spectrum of cyber risks and the necessity for vigilant monitoring and strategic management to safeguard organizations from potential impacts.
- The Cybersecurity Research Institute (2024) highlights the critical nature of monitoring and managing threats like phishing, ransomware attacks, insider threats, APTs, and third-party risks to ensure operational continuity and data protection.
- Martinez, S., & Johnson, M. (2022) stress the significance of early detection and mitigation of insider threats to prevent unauthorized disclosure of critical information and intellectual property theft within organizations.
- Andress, J & Winterfeld, S. (2014) Cyber Warfare, Techniques, Tactics and Tools for Security Practitioners. Second

- Smith, A., & Brown, C. (2023) highlight the critical importance of robust cyber risk monitoring and management in the face of evolving cyber threats, as demonstrated by high-profile events like the SolarWinds cyberattack.

- The Cybersecurity Institute (2024) emphasizes the significance of a Cyber Risk Monitoring Plan (CRMP) as a foundational framework for proactively identifying, assessing, and managing cyber risks to protect digital assets and operations.

- Johnson, T., & Lee, D. (2022) provide insights from case studies such as the Equifax data breach and the WannaCry ransomware attack, showcasing the practical benefits of timely vulnerability patching and effective incident response plans in minimizing cyber incidents through a well-executed CRMP.